Springer Theses

Recognizing Outstanding Ph.D. Research

For further volumes:
http://www.springer.com/series/8790

Aims and Scope

The series "Springer Theses" brings together a selection of the very best Ph.D. theses from around the world and across the physical sciences. Nominated and endorsed by two recognized specialists, each published volume has been selected for its scientific excellence and the high impact of its contents for the pertinent field of research. For greater accessibility to non-specialists, the published versions include an extended introduction, as well as a foreword by the student's supervisor explaining the special relevance of the work for the field. As a whole, the series will provide a valuable resource both for newcomers to the research fields described, and for other scientists seeking detailed background information on special questions. Finally, it provides an accredited documentation of the valuable contributions made by today's younger generation of scientists.

Theses are accepted into the series by invited nomination only and must fulfill all of the following criteria

- They must be written in good English.
- The topic should fall within the confines of Chemistry, Physics, Earth Sciences, Engineering and related interdisciplinary fields such as Materials, Nanoscience, Chemical Engineering, Complex Systems and Biophysics.
- The work reported in the thesis must represent a significant scientific advance.
- If the thesis includes previously published material, permission to reproduce this must be gained from the respective copyright holder.
- They must have been examined and passed during the 12 months prior to nomination.
- Each thesis should include a foreword by the supervisor outlining the significance of its content.
- The theses should have a clearly defined structure including an introduction accessible to scientists not expert in that particular field.

Fangjun Wang

Applications of Monolithic Column and Isotope Dimethylation Labeling in Shotgun Proteome Analysis

Doctoral Thesis accepted by
Dalian Institute of Chemical Physics, Chinese Academy of Science, China

Author
Dr. Fangjun Wang
Dalian Institute of Chemical Physics
Chinese Academy of Science
Dalian
People's Republic of China

Supervisor
Prof. Hanfa Zou
Dalian Institute of Chemical Physics
Chinese Academy of Science
Dalian
People's Republic of China

ISSN 2190-5053 ISSN 2190-5061 (electronic)
ISBN 978-3-662-52432-9 ISBN 978-3-642-42008-5 (eBook)
DOI 10.1007/978-3-642-42008-5
Springer Heidelberg New York Dordrecht London

Printed on acid-free paper

Springer is part of Springer Science+Business Media (www.springer.com)

Parts of this thesis have been published in the following journals:

F. Wang, J. Dong, X. Jiang, M. Ye, H. Zou Anal. Chem. 2007, 79, 6599–606.
F. Wang, X. Jiang, S. Feng, R. Tian, G. Han, H. Liu, M. Ye, H. Zou J. Chromatogr. A 2007, 1171, 56–62.
F. Wang, J. Dong, M. Ye, X. Jiang, R. Wu, H. Zou J. Proteome Res. 2008, 7, 306–10.
F. Wang, M. Ye, J. Dong, R. Tian, L. Hu, G. Han, X. Jiang, R. Wu, H. Zou J. Sep. Sci. 2008, 31, 2589–97.
F. Wang, J. Dong, M. Ye, R. Wu, H. Zou Anal. Chim. Acta 2009, 652, 324–30.
F. Wang, J. Dong, M. Ye, R. Wu, H. Zou J. Chromatogr. A 2009, 1216, 3887–94.
F. Wang, R. Chen, J. Zhu, D. Sun, C. Song, Y. Wu, M. Ye, L. Wang, H. Zou Anal. Chem. 2010, 82, 3007–15.
C. Song, F. Wang, M. Ye, K. Cheng, R. Chen, J. Zhu, Y. Tan, H. Wang, D. Figeys, H. Zou Anal. Chem. 2011, 83, 7755–62.
F. Wang, A. P. Blanchard, F. Elisma, M. Granger, H. Xu, S. A. Bennett, D. Figeys, H. Zou Proteomics 2013, 13, 1292–305.

Supervisor's Foreword

Protein science is the core of modern biological research as all of the biological functions are specifically performed by proteins. Proteomics deals with all of the expressed proteins and protein posttranslational modifications (PTMs) within a living organism. Since the past decade, shotgun proteomics based on the identification and quantification of proteolytic peptides has been greatly developed and attracts widespread concern from the scientific community. Capillary liquid chromatography coupled with mass spectrometry (LC-MS) is the central instrument of shotgun proteomics. The performance of LC separation and related sample preparation methods, such as chemical isotope labeling, is essential to the sensitivity, reproducibility, throughput, accuracy, and coverage of proteome analysis. Therefore, numerous efforts have been made to develop new technologies and strategies for high-efficient proteome analysis in recent years.

This Ph.D. thesis focuses on new monolithic columns and isotope dimethyl labeling strategies and their applications in high-efficient proteome analysis. The phosphate strong cation exchange (SCX) monolithic column with high permeability and high sample loading capacity was developed to improve the fractionating resolution of online multidimensional LC separation. The hydrophobic C12 monolithic column was also developed as LC separation column, electrospray ionization emitter, and monolithic frit to improve the LC separation performance in different applications. Moreover, online isotope labeling was further combined with online multidimensional LC-MS/MS analysis to perform all of the procedures of proteome quantification automatically. All of these technologies and strategies exhibit great potential in high-efficient proteome analysis of different biological samples, such as hepatocellular carcinoma (HCC) liver tissues and hippocampus of model mice with Alzheimer's disease (AD).

Part of this Ph.D. thesis was carried out at University of Ottawa under the supervision of Dr. Daniel Figeys to investigate the phosphoproteome changes during the AD progression of mice model. This thesis demonstrates the new technologies and strategies in LC separation and sample preparation, which play important roles in high-efficient proteome and proteome PTMs analyses.

Dalian, August 2013 Prof. Hanfa Zou

Acknowledgments

This thesis was guided by Prof. Hanfa Zou, and I am extremely grateful for his great help during the time of research and the writing of this thesis. I would like to express my sincere gratitude to Prof. Zou for the continuous support of my Ph.D. study and research, for his motivation, enthusiasm, and immense knowledge. I am heartily thankful to Prof. Figeys for his patience, encouragement, and broad research vision; and for his guidance and support of my research in Ottawa.

Great appreciation is owed to Prof. Mingliang Ye for his selfless help and guidance during the entire research period. I would also like to thank Prof. Ren'an Wu, Prof. Steffany Bennett, Dr. Xiaogang Jiang, Dr. Jing Dong, Dr. Ruijun Tian, Dr. Jean-Philippe Lambert, Dr. Hu Zhou, Dr. Zhibin Ning, Dr. Shuai Wang, Dr. Ted Wright, and other members in Prof. Zou and Prof. Figeys's labs for their encouragement, insightful comments, and kind help in my research.

Finally, I would like to thank my family: my parents, my wife Yanxia Qi, and my brother Fanghua Wang for their great support throughout my life.

August 2013 Fangjun Wang

Contents

Chapter 1
Introduction

Although the whole human genome was sequenced in 2001 and high-throughput genome sequencing technologies were greatly developed in recent years, our understanding of most of the physiological and pathological processes in biological systems is still limited. As all the biological functions are performed by proteins, it is essential to investigate the relationship of the protein expression level with different biological processes. Proteomics is aimed to elucidate all the proteins expressed within a biological system, which is much more complex than genomics for the reason that proteins are constituted by more than 20 types of amino acids and the expressed proteins can be dynamically modified by more than 300 types of post-translational modifications, such as phosphorylation, glycosylation, acetylation, and methylation. [1–3].

In current bottom-up proteomic strategy, the proteins are usually digested at first, then the peptides mixture is analyzed by liquid chromatography coupled with tandem mass spectrometry (LC-MS/MS), which is also termed as shotgun proteomics (Fig. 1.1a). Thousands of proteins could be identified and quantified in just one LC-MS/MS experiment due to the high separation capability of LC and high identification capability of MS. However, it is still impossible to detect all of the expressed proteins within a complex biological system due to the extremely high complexity and high dynamic range of protein sample [4]. Therefore, new technologies in high efficient protein sample preparation, high-resolution protein and peptide separation, and high-speed and high-resolution MS detection have been greatly developed in the past decade to increase the detection coverage, sensitivity, accuracy, and throughput of proteome analysis [5–7]. Currently, proteomics has already been widely applied in different areas of biological studies, such as biomarkers and drug targets discovery, protein identification and quantification, protein localization and interaction, protein PTMs and related biological function, and so on (Fig. 1.1b).

Among different LC separation modes, such as reversed phase (RP), strong-cation exchange (SCX), strong-anion exchange (SAX), etc., RP binary gradient LC separation exhibits the highest separation capability and the best compatibility to electrospray ionization (ESI) MS detection. As mass spectrometer is a concentration-dependent

F. Wang, *Applications of Monolithic Column and Isotope Dimethylation Labeling in Shotgun Proteome Analysis*, Springer Theses, DOI: 10.1007/978-3-642-42008-5_1,

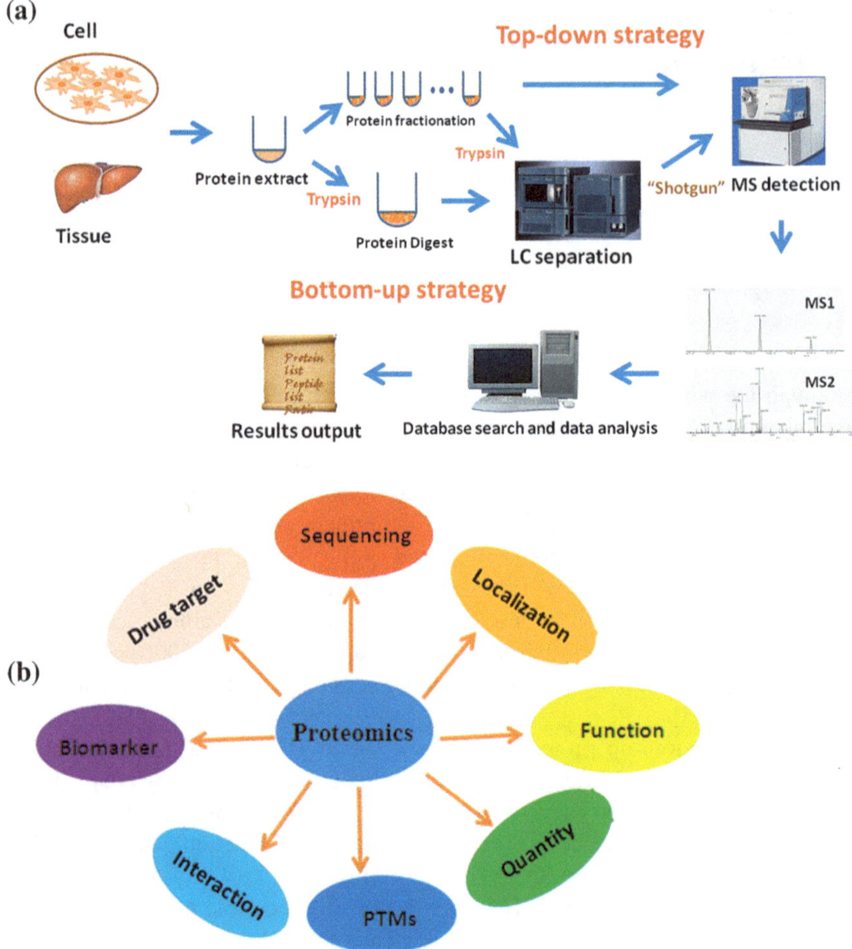

Fig. 1.1 The flow diagram of *top-down* and *bottom-up* (shotgun) proteomics (**a**), and the different types of applications of proteomics (**b**)

detector, decrease in the dimension of LC separation column and the LC separation flow rate can greatly increase the MS detection sensitivity. Usually, the inner diameter (i.d.) of LC separation capillary column is 50–100 μm, and the optimal LC separation flow rate is 100–300 nL/min. Thus, it is difficult to directly inject the protein sample (usually 10–100 μL) onto the separation capillary column. Trap column with larger i.d. (100–250 μm) and shorter column length is usually utilized for fast sample injection (2–20 μL/min) and RP C18 materials are most commonly packed. The dead volume between the RP trap column and RP separation column will significantly decrease the separation performance of LC system. Therefore, Chap. 2 systematically investigates the influence of this type of dead volume on the LC separation performance when C18, C8, and sulfonate SCX trap columns are utilized. We demonstrated that SCX trap column can eliminate the influence of dead volume. On the other hand, online two-dimensional LC separation (2D-LC) can be feasibly achieved due to the orthogonality between SCX trap column and RP separation column, which can improve the detected proteome coverage [8]. Further, to increase the sample loading capacity of the SCX trap column, a 7 cm × 150 μm i.d. phosphate SCX monolithic column was prepared and applied for high-speed sample injection. As the phosphate SCX monolith has higher permeability and sample loading capacity, the sample injection flow rate as well as the resolution in 2D-LC separation were both significantly increased [9].

Monolithic column is considered as the new generation of LC separation column due to its high permeability, fast mass transfer rate, high surface area, high sample adsorption capacity, easy preparation, and versatile surface properties. Different types of monolithic columns have been already developed and widely applied in capillary electrochromatography (CEC) and capillary liquid chromatography (CLC) for high efficient separation of varies samples. However, few reports applied monolithic columns in nanoflow LC-MS/MS for proteome analysis. Therefore, Chap. 3 optimizes the preparation of the C12 organic monolithic column with rigid methacrylate polymer structure to improve its separation performance for digested peptides mixture. The LC separation conditions, such as column length and binary separation gradient were also investigated [10]. The Multidimensional Protein Identification Technology (MudPIT) is widely applied in proteome analyses to increase the detection of proteome coverage by packing RP and SCX materials into a single capillary column in series. Therefore, 10-cm-long phosphate SCX monolith and 65-cm-long C12 RP monolith were also prepared within a single 100-μm-i.d. capillary column, and the dead volume between these two monoliths was eliminated. This SCX-RP biphasic monolithic column exhibited excellent permeability and was successfully applied for online 2D-LC-MS/MS analysis to improve the coverage of proteome analysis [11]. In addition to being used as a LC separation column, the C12 RP monolithic column was also used to fabricate monolithic frit for easy packing of smaller LC stationary materials [12] and ESI emitter for eliminating the post-LC separation column dead volume [13]. And the LC separation performance and proteome coverage of proteome analyses were significantly increased in both cases.

Characterization of the expressed proteins within a biological system by qualitative proteome analysis is usually not enough to understand most of the biological processes. And most of the physiological and pathological processes are directly related to different proteins expression levels as well as different PTMs. Therefore, quantitative proteome analysis plays a more important role to bridge the crack between the expression levels of proteins and protein PTMs and corresponding biological functions. Current technologies for proteome quantification can be divided into two types. The first type is label-free strategy, which utilizes the extracted peak intensity (XIC) in MS detecting chromatogram for corresponding peptides quantification. The label-free strategy is easy to be realized using multiple LC-MS/MS analyses for comparative samples. Further, high sensitivity and high reproducibility can be both achieved in multiple reaction monitoring (MRM) detection mode by selectively extracting the peak intensity of specific fragment in MS/MS for label-free quantification. However, the variations within LC separation, ESI, and so on among comparative samples analyses will compromise the quantification accuracy in all label-free strategies. The other type is stable isotope labeling strategy, which introduces mass differentiate isotope tags into comparative protein samples and the relative quantity of proteins can be obtained in the corresponding MS full scan in proteome analysis of the mixed protein samples. As the comparative samples are mixed together for LC-MS/MS analysis simultaneously, the variations in LC separation and MS detection can be eliminated, which is helpful to improve the accuracy of proteome quantification. The mass differentiate isotope tags can be incorporated into protein samples by in vivo metabolic methods, such as Stable Isotope Labeling by Amino acids in Cell culture (SILAC), or by in vitro chemical reactions, such as Isobaric Tags for Relative and Absolute Quantification (iTRAQ) and dimethyl isotope labeling strategies. The in vivo metabolic labeling methods are usually applied for cell line samples labeling. For animal models and clinical samples, in vitro chemical labeling strategies are more feasibly applied. Dimethyl isotope labeling strategy specifically labels the primary amino groups on lysine residues and N-terms of digested peptides with extra high efficiency and low cost, and it has been already widely applied for quantitative proteome or proteome PTMs analyses of different types of biological samples, such as cell lines, animal tissues, and clinical samples.

There are two problems in current isotope labeling strategies for quantitative proteome analyses. The first one is low sensitivity. After protein digestion, a lot of procedures for sample preparation are needed before LC-MS/MS analysis, such as desalting, lyophilization, redissolve, isotopelabeling, re-desalting, and re-lyophilization. Sample loss and contamination are easily introduced at each of these sample preparation procedures, which significantly decrease the sensitivity and accuracy of proteome quantification. Second, there is lack of reliable criteria to estimate the quantification accuracy. Most of the reported works applied SDS-PAGE and Western Blot (WB) to validate some of the selected proteins within the results of proteome quantification. However, WB is time- and labor-consuming and it is impossible to validate thousands of proteins within the results of proteome quantification. Chapter 4 investigates the combination of online dimethyl isotope labeling with online multidimensional LC separation. A biphasic trap column with 7-cm-long phosphate SCX monolith and 7-cm-long C18 particulate materials was utilized for fast sample injection, online isotope labeling, online sample purification, and online 2D-LC separation. As all the sample preparation procedures were automatically performed within the 200-μm-i.d. capillary trap column, the sample loss and contamination were significantly reduced, which greatly increased the quantification sensitivity, accuracy, and proteome coverage [14]. It was also observed that replicate proteome quantification can significantly improve the quantification accuracy after controlling the coefficient of variation (CV). However, it is sample and time-consuming to perform replicate proteome quantification, which decreases the analysis throughput obviously. Moreover, the sample amount is limited for some clinical or animal model samples, such as only 1–2 mg proteins can be obtained from the hippocampus of one mouse. Therefore, a pseudo triplex dimethyl isotope labeling strategy was further developed to achieve replicate proteome quantification in just one LC-MS/MS experiment. Briefly, two identical control samples are labeled with light (L) and heavy (H) isotopes respectively, while another comparative sample is labeled with a medium (M) isotope. Thus, two replicate analyses can be achieved in just one LC-MS/MS experiment as the M/L and H/L both reveal the relative ratio between comparative and control samples. It was demonstrated that high proteome coverage and high quantification accuracy can be both obtained by this pseudo triplex dimethyl labeling strategy in proteome and phosphoproteome analyses [15]. Then, this pseudo triplex isotope labeling strategy was successfully applied to investigate the different phosphoproteome expression of HCC and normal liver tissues [15] as well as the dynamics of phosphoproteome expression of early onset mouse model (TgCRND8) of Alzheimer's disease (AD) [16].

Overall, this dissertation focuses on improving the performance of qualitative and quantitative proteome analyses by LC-MS/MS. And new technologies in monolithic column and stable isotope labeling are feasibly developed to improve the throughput, sensitivity, reproducibility, accuracy, and proteome coverage in large-scale proteome analyses. In particular, the phosphate SCX monolithic column exhibits much higher permeability and sample binding capacity than conventional sulfonate SCX column, which significantly improve the sample injection flow rate and SCX online fractionating resolution. This phosphate SCX monolithic column is widely applied in proteome analysis of many types of biological samples, such as oleaginous yeast Lipomyces starkeyi and engineered tendon in our laboratory [17–19]. Further, we combine both online dimethyl isotope labeling and 2D-LC-MS/MS into an automated system, which greatly improve the sensitivity, accuracy, and coverage of proteome quantification. In addition, the pseudo triplex dimethyl isotope labeling strategy improves the accuracy and coverage of proteome quantification by using limited amount of protein samples. Therefore, it can be seen that new technologies are especially important for proteome analyses of complex biological samples. There is still a long way to go for proteomics to become a mature tool as genomics. A lot of challenges within this process are actually analytical chemistry problems, such as the low sensitivity, low accuracy, low reproducibility, and low throughput. We believe much work still needs to be done to promote the development of proteomics, particularly new analytical technologies in both high performance LC separation and protein sample preparation. The paradigms provided in this dissertation add some knowledge on how to improve the performance of proteome analyses by using new monolithic column and stable isotope labeling strategies.

References

1. Pandey A, Mann M (2000) Proteomics to study genes and genomes. Nature 405:837–846
2. Aebersold R, Mann M (2003) Mass spectrometry-based proteomics. Nature 422:198–207
3. Nilsson T, Mann M, Aebersold R, Yates JR 3rd, Bairoch A, Bergeron JJ (2010) Mass spectrometry in high-throughput proteomics: ready for the big time. Nat Methods 7:681–685

4. Cox J, Mann M (2008) MaxQuant enables high peptide identification rates, individualized p.p.b.-range mass accuracies and proteome-wide protein quantification. Nat Biotechnol 26:1367–1372
5. Wisniewski JR, Zougman A, Nagaraj N, Mann M (2009) Universal sample preparation method for proteome analysis. Nat Methods 6:359–362
6. Xie C, Ye M, Jiang X, Jin W, Zou H (2006) Octadecylated silica monolith capillary column with integrated nanoelectrospray ionization emitter for highly efficient proteome analysis. Mol Cell Proteomics 5:454–461
7. Mann K, Mann M (2011) In-depth analysis of the chicken egg white proteome using an LTQ Orbitrap Velos. Proteome Sci 9:7
8. Wang F, Jiang X, Feng S, Tian R, Han G, Liu H, Ye M, Zou H (2007) Automated injection of uncleaned samples using a ten-port switching valve and a strong cation-exchange trap column for proteome analysis. J Chromatogr A 1171:56–62
9. Wang F, Dong J, Jiang X, Ye M, Zou H (2007) Capillary trap column with strong cation-exchange monolith for automated shotgun proteome analysis. Anal Chem 79:6599–6606
10. Wang F, Dong J, Ye M, Wu R, Zou H (2009) Improvement of proteome coverage using hydrophobic monolithic columns in shotgun proteome analysis. J Chromatogr A 1216:3887–3894
11. Wang F, Dong J, Ye M, Jiang X, Wu R, Zou H (2008) Online multidimensional separation with biphasic monolithic capillary column for shotgun proteome analysis. J Proteome Res 7:306–310
12. Wang F, Dong J, Ye M, Wu R, Zou H (2009) Integration of monolithic frit into the particulate capillary (IMFPC) column in shotgun proteome analysis. Anal Chim Acta 652:324–330
13. Wang F, Ye M, Dong J, Tian R, Hu L, Han G, Jiang X, Wu R, Zou H (2008) Improvement of performance in label-free quantitative proteome analysis with monolithic electrospray ionization emitter. J Sep Sci 31:2589–2597
14. Wang F, Chen R, Zhu J, Sun D, Song C, Wu Y, Ye M, Wang L, Zou H (2010) A fully automated system with online sample loading, isotope dimethyl labeling and multidimensional separation for high-throughput quantitative proteome analysis. Anal Chem 82:3007–3015
15. Song C, Wang F, Ye M, Cheng K, Chen R, Zhu J, Tan Y, Wang H, Figeys D, Zou H (2011) Improvement of the quantification accuracy and throughput for phosphoproteome analysis by a pseudo triplex stable isotope dimethyl labeling approach. Anal Chem 83:7755–7762
16. Wang F, Blanchard AP, Elisma F, Granger M, Xu H, Bennett SA, Figeys D, Zou H (2013) Phosphoproteome analysis of an early onset mouse model (TgCRND8) of Alzheimer's disease reveals temporal changes in neuronal and glia signaling pathways. Proteomics 13:1292–1305
17. Jiang Y, Liu H, Li H et al (2011) A proteomic analysis of engineered tendon formation under dynamic mechanical loading in vitro. Biomaterials 32:4085–4095
18. Liu H, Zhao X, Wang F, Li Y, Jiang X, Ye M, Zhao ZK, Zou H (2009) Comparative proteomic analysis of Rhodosporidium toruloides during lipid accumulation. Yeast 26:553–566
19. Zhu Z, Zhang S, Liu H et al (2012) A multi-omic map of the lipid-producing yeast Rhodosporidium toruloides. Nat Commun 3:1112

Chapter 2
Online Sample Injection and Multidimensional Chromatography Separation by Using Strong-Cation Exchange Monolithic Column

2.1 Introduction

Nanoflow LC-MS/MS is the central platform in shotgun proteome analysis due to its high sensitivity and high separation capability. The flow rate of nanoflow LC separation is usually 50–300 nL/min. Manual injection of protein sample onto a 50–100 μm inner diameter (i.d.) capillary LC separation column by nitrogen pressure or void capillary is feasibly achieved in previous proteome analyses [1–3]. However, this strategy is time-consuming, labor-intensive, and variations are easily introduced in the manual operations, which significantly decreases the analysis throughput and reproducibility. Automated sample injection by using a trap column with shorter column length and larger inner diameter is the most popular way in nanoflow LC-MS/MS analyses [4]. Briefly, protein samples vary from several to hundreds of microliters are loaded onto the trap column at a high flow rate in short time at first, and after equilibrium the adsorbed peptides are eluted from the trap column to a reversed phase (RP) capillary column for LC separation. There are usually two types of instrument configurations for the automated sample injection in nanoflow LC-MS/MS analyses. In the first type of configuration, the trap column and RP separation capillary column are connected by a nanoflow switching valve. The flow through from trap column is directed to waste or LC separation column during sample injection or separation by the switching valve [5–7]. Proteolytic digest without prior purification could be directly injected and cleaned up online in this case, which is important in proteome analysis due to the low sample loss. However, the void volume introduced by switching valve would seriously degrade the LC separation performance. The other one is vented column system in which trap and separation columns are directly connected via a microcross or microtee with an open/close switching valve [8–11]. A regular 6-port switching valve could be used instead of using nanoflow switching valve in this type of configuration due to the mobile phase for LC separation does not pass though the switching valve. Further, to minimize the void volume introduced by the microcross or microtee, Licklider et al. packed the open space of microcross with C18 particles to reduce void volume [8].

F. Wang, *Applications of Monolithic Column and Isotope Dimethylation Labeling in Shotgun Proteome Analysis*, Springer Theses, DOI: 10.1007/978-3-642-42008-5_2,

Meiring et al. drilled the microtee to 0.6 mm i.d. to fit a single micro sleeve having a V-shaped cut as a waste outlet, and the trap and analytical columns were butt-connected in this modified sleeve [9]. However, these systems need tedious manual operation and the reproducibility is difficult to control.

As described in Chap. 1, proteins within biological systems are extremely complex and more than 300 types of posttranslational modifications (PTMs) can be dynamically occurred. Further, the dynamic expression range of different proteins exceeds six orders of magnitude in cells and ten orders of magnitude in body fluids [12]. For large-scale protein identification, powerful separation strategies coupling to mass spectrometry are usually applied to reduce the complexity of protein analytes. And C18 RP capillary column is most widely used due to its high separation efficiency and good compatibility with ESI MS detection. To improve the LC separation performance of C18 capillary column, longer columns and/or smaller packing materials are usually utilized [7, 13, 14]. However, the operating pressure of these types of capillary columns increases dramatically, and can only be implemented using advanced LC instrumentations with ultra-high operating pressure. Multidimensional LC separation is another alternative to increase the separation capability of the whole LC system, which is typically achieved by offline or online coupling of the two columns with orthogonal retention mechanisms [15–19]. The separation capability of multidimensional LC separation depends on both of the separation performance and orthogonality of the two different dimensions and the total separation peak capacity could improve from ~10^2 to ~10^3 [20, 21]. Among all of these cases, C18 RP chromatography was usually utilized as the first LC dimension that directly coupled to MS, and strong-cation exchange (SCX) chromatography was most widely used as the second dimension for peptides or proteins prefractionation due to its good orthogonality to RP chromatography.

Comparing with offline strategy, online multidimensional LC separation exhibits low sample loss and contamination, and the analysis sensitivity and throughput can be also greatly improved. Multidimensional Protein Identification Technology (MudPIT) is the most widely utilized mode of online multidimensional LC separation, and it is achieved by a biphasic capillary column packed with SCX and RP materials in sequence [1, 2, 22]. Peptides loaded onto the SCX materials are stepwise eluted to RP segment by flushing salt step, and a nanoflow RP LC-MS/MS analysis is flowed with each salt step. This online 2D LC-MS/MS is fully automated and exhibits advantages such as minimal sample loss, no vial contamination, and sample dilution. However, the LC operation pressure limits the packing amount of SCX materials, which compromise the sample loading capacity and SCX online fractionating resolution. In order to increase the sample loading amount for multidimensional separation, overloading of the SCX segment could easily occur. Further, it is time-consuming and labor-intensive to manually load protein sample onto the SCX-RP biphasic particulate column due to the relative low permeability.

Monolithic column is a good alternative to particulate column due to its extremely low back pressure, high surface area, high mass transfer rate, and high sample loading capacity [23–26]. Polymer-based organic monolithic columns with sulfonate SCX stationary phases exhibits advantages such as pH stability, inertness to biomolecules, absence of deleterious effects from silanol, and facility for modification [23, 24]. Various strategies were developed to introduce sulfonate groups into the monolith backbone, such as adsorption of surfactants [27, 28] grafting of pore surface [29], modification of reactive monoliths [30], and copolymerization of crosslinker and monomer containing the function group [31–33]. Copolymerization of crosslinker and functional monomer is the most straightforward strategy and the sample loading capacity can be controlled by adjusting the amount of functional monomer in the polymerizing mixture [34].

As described above, the dead volume resulted from the connections between C18 trap and C18 LC separation capillary columns inevitably decreases the separation performance of the LC system, and further decreasing the dead volume is a technique challenge. Therefore, a good alternative is to develop a dead volume insensitive system for sample injection and nanoflow LC-MS/MS analysis. Instead of using C18 trap column, we found recently that SCX trap column can alleviate the influence of dead volume on LC separation in vented column system [3]. However, proteolytic digest sample must be cleaned before sample injection in order to eliminate the contamination of the LC separation column. In this chapter, the influence of void volume between trap and separation columns on the LC separation performance and proteomic coverage was systematically studied, and it was found that the influence of dead volume (varied from 0 to 5 μL) was nearly eliminated by using SCX trap column. Then an automatic sample injection system was constructed by using 10-port switching valve and SCX trap column, and proteolytic digest containing contaminants could be injected directly without purification. This system allowed fast sample injection at ~2 μL/min. To further improve the sample injection flow rate and the resolution in online fractionation for the SCX trap column, we then developed a phosphate monolithic capillary column by direct copolymerization of a ethylene glycol methacrylate phosphate (EGMP) and bis-acrylamide in a trinary porogenic solvent consisting dimethylsulfoxide, dodecanol and N, N′-dimethylformamide within a 150 μm-i.d. capillary column. When coupled this phosphate monolithic SCX trap column to a C18 packed capillary column with integrated ESI tip, this system exhibited good separation performance as well as high proteomic coverage in both one- and multidimensional nanoflow LC-MS/MS analysis. Due to the extremely low backpressure of phosphate monolithic column, automated sample injection at high flow rate of 40 μL/min could be easily achieved. This type of phosphate monolithic column provided a reliable alternative to the particulate SCX columns for proteome analysis.

2.2 Experimental Results

Capillary LC columns with 50–300 nL/min separation flow rate are widely applied in proteome analysis to increase the MS detection sensitivity as mass spectrometer is a concentration-dependent detector. Due to the extreme complexity and high dynamic range of protein samples, high performance LC separation is essential to large-scale proteome analyses. C18 capillary column is the most widely used one among all the LC separation columns in proteome analyses due to the RP binary gradient separation exhibits higher separation resolution than other separation modes, such as strong-cation exchange (SCX), strong-anion exchange (SAX), and hydrophilic interaction liquid chromatography (HILIC). As C18 trap column with 100–300 μm inner diameter (i.d.) is widely utilized for fast sample injection as described above, the dead volume between C18 trap and LC separation column compromises the performance of LC separation. Therefore, we try to eliminate this influence at first in this chapter.

2.2.1 The Influence of Dead Volume on LC Separation Performance Using Different Types of Trap Columns

We investigated the influence of dead volume in LC-MS/MS systems using C18, C8, and SCX trap columns, respectively. C18 trap column system was studied at first, and the dead volume of 0, 1, 3, and 5 μL between trap and separation columns was investigated, respectively, by using different length of 150 μm-i.d. void capillary (Fig. 2.1a, b). After LC-MS/MS analyses for 1 μg tryptic digest

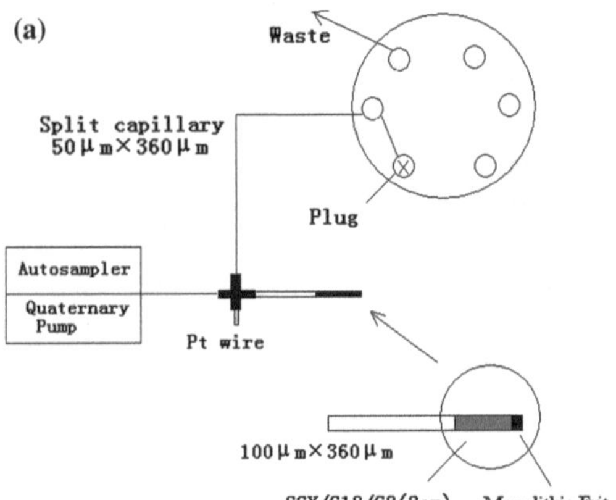

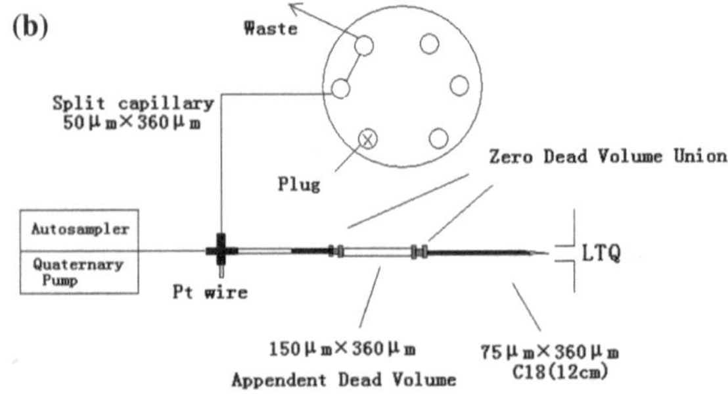

Fig. 2.1 Schematic diagrams of the offline sample injection system to investigate the influence of dead volume on the LC separation performance, **a** loading sample onto trap column; **b** eluting sample onto analytical column and gradient RPLC analysis

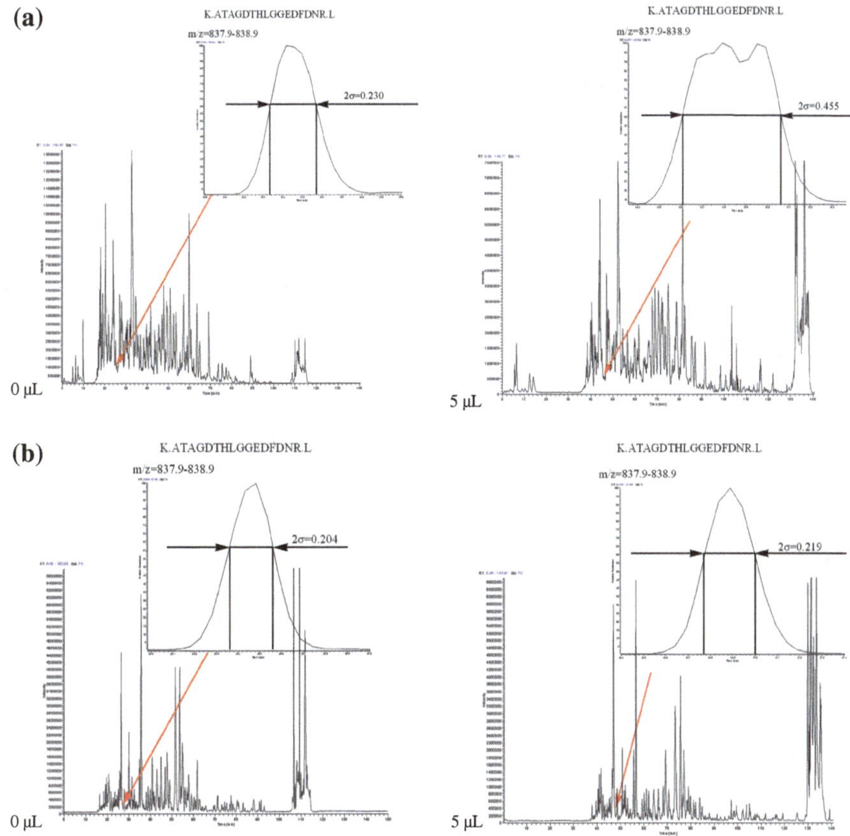

Fig. 2.2 Base peak chromatograms for nanoflow RPLC-MS/MS analysis of 1 μg SPE cleaned yeast protein tryptic digest on the offline sample injection system using **a** C18 and **b** SCX trap columns with void volume of 0 μL and 5 μL between trap and analytical columns

of extracted yeast protein, one identified peptides K.ATAGDTHLGGEDFDNR.L (m/z = 838, +2) with moderate abundance was extracted from each of the chromatogram to estimate the variation of LC separation capacity. As shown in Fig. 2.2a, the peak capacity decreased about 50 % after the dead volume increased from 0 to 5 μL for C18 trap column system. After database searching and controlling false discovery rate (FDR) < 1 %, 544 and 382 unique peptides corresponding to 270 and 195 distinct proteins were positively identified for systems with 0 and 5 μL dead volume, respectively. It could be seen that the peak capacity and peak intensity in LC separation, and the numbers of identified peptides and proteins were all decreased along with the increase of dead volume (Table 2.1 and Fig. 2.3a, b, the peak capacity is defined as $n_c = L/(4\sigma)$, where L is the total time over which the peptides eluted and σ is the average standard deviation of the peaks, 2σ can be obtained as the peak width at 0.613 peak height). Obviously, the

Table 2.1 Peak capacity and intensity for offline sample injection system by using C18, C8, and SCX trap columns with a void volume of 0, 1, 3, and 5 μL between trap and analytical columns as well as manual injection system

Trap Column	Void volume (μL)	2σ	Peak capacity	Peak intensity(10^6)
C18	0	0.230	200	1.82
	1	0.254	181	1.67
	3	0.290	159	1.48
	5	0.455	101	0.50
C8	0	0.242	190	0.88
	1	0.184	250	0.87
	3	*	*	*
	5	0.248	185	1.09
SCX	0	0.204	225	1.15
	1	0.208	221	1.54
	3	0.215	214	1.66
	5	0.219	210	1.03
Manual injection	\	0.189	243	2.96

* K.ATAGDTHLGGEDFDNR.L has not been detected in this analysis

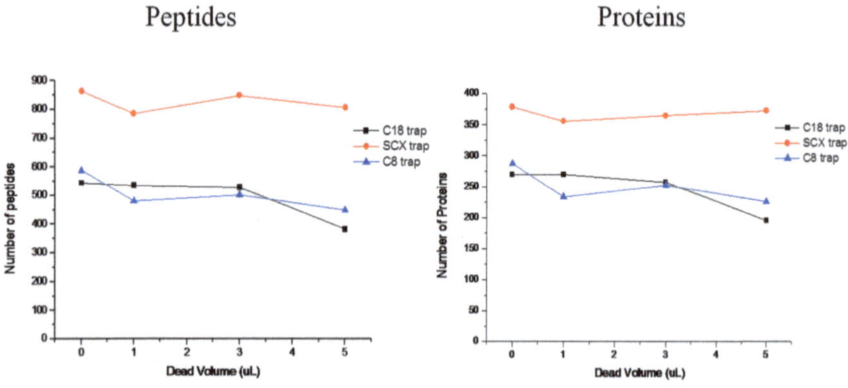

Fig. 2.3 Effect of dead volume (between trap and LC separation columns) on the number of unique peptides and distinct proteins identified on offline sample injection system using C18, C8, and SCX trap columns (Redrawn with permission from Ref. [35]. Copyright 2007 Elsevier)

increase of dead volume between C18 trap and LC separation columns seriously decreased the LC separation capability in RP binary gradient separation, and the numbers of identified peptides and proteins both decreased ~30 % in proteome analysis when the dead volume increased from 0 to 5 μL. Then, the C8 trap column was investigated in the same way as described above. Similar to C18 trap column, the numbers of identified peptides and proteins were also decreased along with the increasing of dead volume. Interestingly, the proteome coverage of C8 trap column system is inferior to C18 trap column system when the dead volume is 0 μL, but better proteome coverage can be obtained for C8 trap column system when the dead volume increases to 5 μL (Fig. 2.3a, b). This can be explained by the refocusing effect of C8 trap column. As the hydrophobicity of C8 is lower than C18 materials, peptides eluted from C8 trap column will be refocused onto the head of C18 LC separation column, which can slightly alleviate the influence of dead volume on the LC separation performance as well as proteome coverage [9].

Finally, the SCX trap column was utilized and the dead volumes 0, 1, 3, and 5 μL were also investigated, respectively. In contrast to C18 and C8 trap columns, the peak capacity and intensity in LC separation and the numbers of identified peptides and proteins were almost not changed in SCX trap column system even if the dead volume was increased from 0 to 5 μL (Figs. 2.2b, 2.3a, b, and Table 2.1). Obviously, better performances were obtained for SCX trap column system in both LC separation and proteome coverage. A total of 379 and 373 distinct proteins were successfully identified for SCX trap column system with 0 and 5 μL dead volumes, which increased 40 and 91 % compared with C18 trap column system at similar conditions, respectively. Similar results were also obtained for the identified unique peptides. Therefore, the numbers of identified unique peptides and proteins in the system using SCX trap column was hardly affected by the increase of dead volume (Fig. 2.3a, b). That might be explained by the migration behavior of analytes in this system. In the system using SCX trap column, the analytes migration behavior was different from C18 and C8 trap column systems. After sample was injected onto the SCX trap column, the peptides bounded on SCX resin with electrostatic interaction were eluted onto the C18 separation column by flushing with 500 mM NH4Ac buffer (pH 3) for 10 min. Although dilution and mixing would occur in the void space when the peptides were eluted, all the peptides were re-enriched again on the front of C18 analytical column by hydrophobic interaction. When the binary gradient started for RP separation, there were no peptides eluted from SCX trap column to participate in LC separation. And the separation of peptides was only performed on the C18 separation column. Though gradient delay due to the dead volume would happen, dilution and mixing of the gradient profile would less affect the LC separation performance on C18 column (Fig. 2.2). However, in the systems using C18 and C8 trap columns, the RP binary gradient separation is started from the C18 or C8 trap column, the eluted analytes from the trap column as well as the RP gradient profile are seriously mixed and diluted within the dead volume of connecting void capillary, microtee, microcross, and union, which seriously decreases the performance of both LC separation and proteome identification (Fig. 2.3 and Table 2.1). In conclusion, the tolerance to dead volume between trap and separation columns is SCX $\gg$ C8 > C18, and SCX trap column exhibits superior performance in both LC separation and proteome analyses.

Manual injecting of protein sample directly onto the LC C18 separation column by using a void capillary can eliminate the dead volume as no trap column is needed. Thus, we also utilized manual injection of 1 μg yeast protein digest for comparison. After nanoflow LC-MS/MS analysis, the moderated intensity peptide of K.ATAGDTHLGGEDFDNR.L ($m/z = 838, +2$) was also extracted from the chromatogram, and the peak capacity and intensity were 243 and 2.96×10^6, respectively, which is better than all of the vented systems using trap columns (Table 2.1). After database searching, 929 unique peptides from 392 distinct proteins were identified,

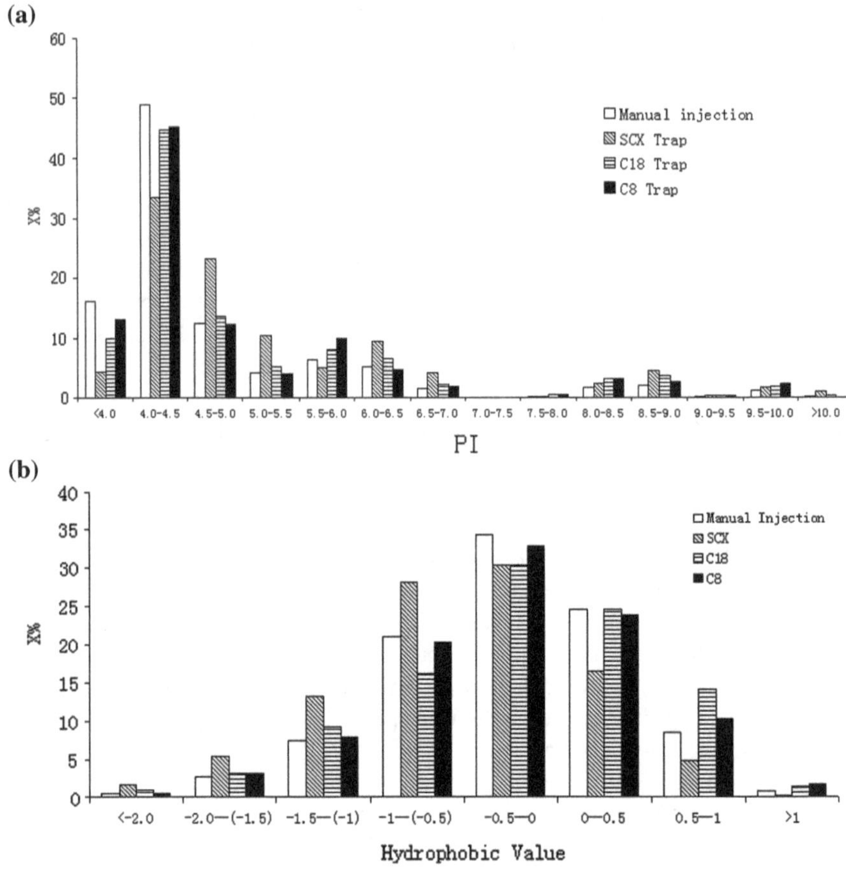

Fig. 2.4 Distributions of **a** pI value and **b** hydrophobicity (GRAVY) for the identified peptides by nanoflow LC-MS/MS analyses in manual injection, offline sample injection by using C18, C8, and SCX trap columns with void volume of 5 μL between trap and analytical columns (Redrawn with permission from Ref. [35]. Copyright 2007 Elsevier)

which is also better than all of the trap column systems (Fig. 2.3a, b). Therefore, the performance of both the LC separation and proteome analysis can be increased if no trap column was applied. However, manual injection is tedious and time-consuming, which decrease the throughput and reproducibility of proteome analyses. Among all of the trap columns, SCX trap column exhibits the best performance, and the numbers of identified proteins and peptides are also comparable to manual injection. Then, the pI value and hydrophobicity of all the identified peptides by different sample injection methods were calculated. It could be seen that the C18 and C8 trap columns systems could identify more peptides in low pI value range (these peptides is difficult to be charged in 0.1 % formic acid (FA) separation buffer) (Fig. 2.4a), and SCX trap column can identify more peptides in low hydrophobicity (Fig. 2.4b), which are consistent with the retention mechanism of different trap columns.

2.2.2 Online Sample Injection of Uncleaned Protein Sample SCX-RP 2D LC-MS/MS Analysis by SCX Trap Column

As described above, the SCX trap column has the highest tolerance to dead volume between trap and LC separation columns. Thus, we developed an online sample injection system for automated sample injection and LC-MS/MS analysis. The SCX trap column was connected with the C18 LC separation column by microfilter, nanoflow 10-ports switching valve, and microtee (Fig. 2.5). In this system, the uncleaned protein sample can be directly injected onto the trap column and the contaminants, such as salt, detergent, etc., can be online removed by the switching valve, which is important to decrease the sample loss in purification of limited amount of sample. 1 μg uncleaned tryptic digest of yeast protein extract was automatically injected onto the SCX column and equilibrated by 0.1 % FA aqueous solution for purification. Then, the peptides were transferred to the C18 separation column by a 10 min elution of 1,000 mM NH_4AC solution (pH 2.7) and equilibrated again by 0.1 % FA aqueous solution before nanoflow LC-MS/MS analysis. The efficient LC separation window was started from 30 min in the chromatogram (Fig. 2.6a), which implied the dead volume between the trap and separation column was about 3 μL as the separation window in manual injection was started from 15 min and the flow rate is 200 nL/min. Further,

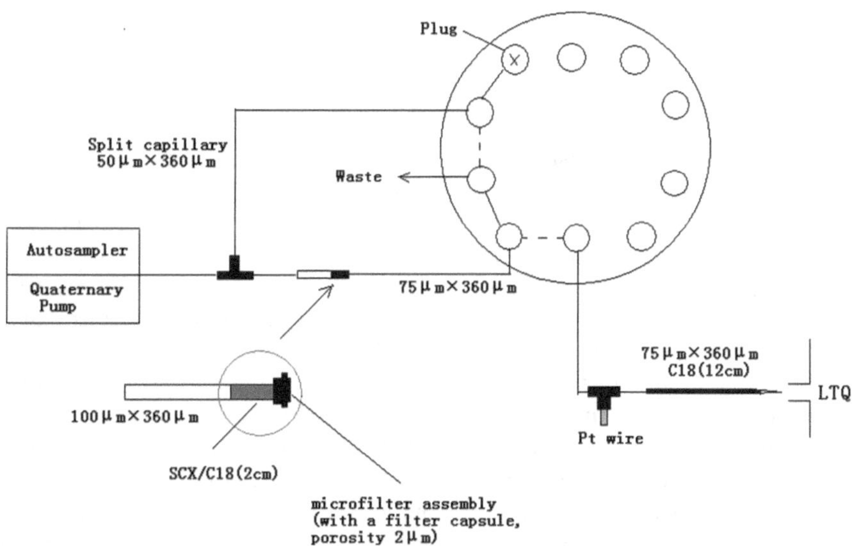

Fig. 2.5 Schematic diagram of the system for injection of the uncleaned protein sample using 10-ports switching valve. *Solid line* loading sample onto trap column; *dashed line* eluting sample to analytical column and gradient RPLC analysis (Reprinted with permission from Ref. [35]. Copyright 2007 Elsevier)

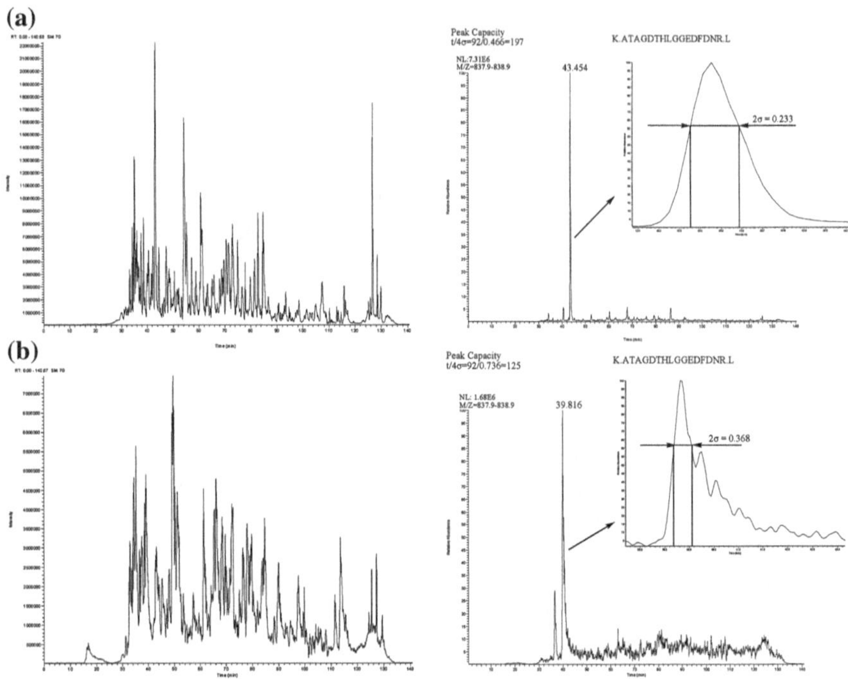

Fig. 2.6 Base peak chromatogram for directly nanoflow LC-MS/MS analyses of 1 μg tryptic digest of yeast proteins with online sample injection system using **a** SCX and **b** C18 trap column. The peak profile using extracted ion chromatogram corresponding to individual peptide was used for calculating peak capacity of this system (Redrawn with permission from Ref. [35]. Copyright 2007 Elsevier)

as the dead volume is consisted by microfilter, switching valve, microtee, and connecting capillary, which is much more complex than the 3 μL dead volume introduced just by void capillary as described above in offline sample injection. Therefore, comparing to the offline sample injection system using SCX trap column, the peak width (2σ, at 0.613 peak height) of the moderated intensity peptide of K.ATAGDTHLGGEDFDNR.L (m/z = 838, +2) was wider in this online sample injection system (Table 2.1 and Fig. 2.6a). Finally, 1,077 unique peptides corresponding to 403 distinct proteins were identified, which were better than the results obtained by manual injection of 1 μg SPE cleaned protein samples. Therefore, online sample cleaning by switching valve can decrease the sample loss and increase the proteome coverage, which is especially benefit for limited amount samples analyses. C18 trap column was also applied for comparison, and the detected peptide peak was seriously expanded and branched. After dataset searching, only 405 unique peptides corresponding to 195 proteins were identified. These results also demonstrated that the SCX trap column has much higher tolerance to dead volume between trap and C18 LC separation columns.

Table 2.2 Run-to-run retention reproducibility[a] of four consecutive analyses of yeast protein tryptic digest by online sample injection system using SCX trap column

Mass	Retention time					RSD (%)
	Run 1	Run 2	Run 3	Run 4	Ave.	
1,424.66	29.82	30.99	33.62	31.73	31.54	4.4
2,011.02	35.72	37.14	39.57	37.82	37.56	3.7
2,294.02	42.85	44.20	46.50	45.00	44.64	3.0
3,000.40	51.38	52.22	53.92	53.14	52.66	1.8
1,427.68	57.86	59.18	61.12	60.34	59.62	2.1
2,058.02	61.66	62.89	64.84	64.13	63.38	1.9
1,815.98	66.88	67.92	70.12	69.63	68.64	1.9
1,199.67	71.88	73.24	75.15	74.44	55.37	2.5
2,418.13	78.22	79.27	81.06	80.63	79.80	1.4
1,278.66	85.36	86.95	88.96	88.26	87.38	1.6
1,719.87	91.72	93.99	95.67	94.94	94.08	1.6
2,143.13	99.58	104.18	106.03	105.12	103.73	2.4
2,577.27	102.38	106.36	108.08	107.42	106.06	2.1
2,168.18	109.53	111.70	112.61	112.09	111.48	1.1
1,821.92	122.54	124.45	124.68	124.87	124.13	0.8

[a]Peptides are randomly selected across the elution, sample loading rate was 2 μL/min, and other conditions are the same as Fig. 2.6a

Then, the reproducibility of this automated nanoflow LC-MS/MS system with SCX trap column was further evaluated, and four replicated analyses were performed. The retention time of 15 peptides was obtained from corresponding elution profiles in chromatograms and the relative standard deviations (RSDs) were all <5 % (Table 2.2). After database searching, the average numbers of identified unique peptides and proteins were 1,053 and 403, and the RSDs were 6.0 and 2.9 %, respectively, for four-time replication.

Although RP binary gradient separation exhibits better separation capability than other LC separation modes, the separation capability is still not enough due to the extreme complexity of protein samples as described in introduction of this chapter. Therefore, multidimensional LC separations by combing LC columns with different separation mechanisms are rapidly developed for proteome analyses in recent years. The most common combination in multidimensional LC separation is SCX and RP due to their good orthogonality. Online SCX-RP two-dimensional LC separation is applied for qualitative and quantitative proteome analyses due to its high separation performance and high detection sensitivity. In our automated sample injection system by using SCX trap column, online two-dimensional LC separation can be easily achieved by online stepwise elution of different concentration of NH_4AC buffer (pH 2.7) as described in Experimental Section. Therefore, this automated online two-dimensional LC separation system was utilized for high efficient separation of 20 μg tryptic digest of yeast protein extract. The digested peptides were online cleaned by the SCX trap column, followed by 13 steps online fractionations, and one nanoflow LC-MS/MS analysis was applied after each fractionation. Finally, 2,388 unique peptides and 870 proteins were successfully identified. Comparison to one-dimensional RP LC separation, the proteome coverage increased 120 %, and the distribution of peptides pI values and hydrophobicity was more symmetrical (data not shown). Therefore, for complex protein sample analysis, online SCX-RP two-dimensional separation can greatly improve the proteome coverage and detection sensitivity.

2.2.3 The Preparation and Characterization of Phosphate SCX Monolithic Column

In the previous works, the particulate SCX column packed with sulfonate silica gels was utilized for online sample injection and SCX-RP 2D LC-MS/MS analyses. One major disadvantage of packed SCX column is the high back pressure, which greatly limits the packed length and sample loading amount. And overloading of protein sample is easily happened in the system using packed SCX column for sample injection, which will decrease the resolution of SCX chromatography in online fractionation. Therefore, increase of the sample loading capacity of SCX trap column is essential to increase both the performance of SCX-RP online two-dimensional LC separation and the identified proteome coverage. Comparing with packed capillary column, capillary monolithic column has higher permeability and specific surface area, and has been already widely applied in capillary electrochromatography (CEC) analyses of different samples. However, the application of monolithic column in proteome analyses of complex biological samples is still limited, especially for SCX monolithic column. SCX monolithic columns, prepared by various sulfonate functional monomers, have been prepared and applied in CEC separation. However, the sulfonate SCX monolithic columns always exhibit extra swelling or shrinking during the change of different LC separation buffer, such as ACN and aqueous solutions [30–32]. Furthermore, it is observed that the conventional SCX monolithic columns always have part of RP retention mechanism besides SCX retention mechanism [34, 36]. Also this type of SCX monolithic column with mix retention modes sometimes have better separation performance for specific analytes, the orthogonality to RP separation will decrease in multidimensional LC separation. Therefore, it is important to develop new type of SCX monolithic column with pure SCX retention mechanism, high stability, and high sample loading capacity to further increase the LC separation performance of SCX-RP 2D LC system for complex protein sample analyses. In our previous work, a phosphate SCX monolithic capillary column was prepared by direct copolymerization of a EGMP and bis-acrylamide in a trinary porogenic solvent consisting dimethylsulfoxide, dodecanol and N, N'-dimethylformamide. It was demonstrated that this phosphate monolithic column was able to specifically isolate phosphopeptides after Zr^{4+} was loaded onto the column [37]. Therefore, we also investigated if this phosphate SCX monolithic column can be used as trap column for fast sample injection and SCX-RP 2D LC separation.

In order to increase the sample loading capability as well as the sample loading rate, we synthesized the phosphate monolith in confined 150-μm-i.d. capillary column, and the obtained monolith was characterized by scanning electron micrograph as shown in Fig. 2.7. It can be seen that the monolithic bed closely linked to the pretreated capillary wall and macropores are feasibly formed by the trinary

Fig. 2.7 Scanning electron microscopy photographs of phosphate monolith inside a capillary with 150 μm-i.d. at magnification of 500× and 5,000× (Reprinted with permission from Ref. [38]. Copyright 2007 American Chemical Society)

Table 2.3 Permeability of different trap columns

Column (150 μm i.d.)	Flushing solution	Viscosity, η (cP)[a]	Back pressure, $\triangle P$ (psi)	Flow rate, F (μL/min)	Permeability, k ($\times 10^{-14}$ m^2)[b]
Phosphate monolithic column (7 cm)	Acetonitrile	0.369	594.5	80	47.5
	Water	0.890	1,290.5	40	26.4
Particulate SCX column (1 cm)[c]	Acetonitrile	0.369	1,483.4	80	2.7
	Water	0.890	2,128.6	40	2.3
Particulate C18 column (1 cm)[c]	Acetonitrile	0.369	1,348.5	80	3.0
	Water	0.890	1,679.1	40	2.9

[a]Viscosity data were from Ref. [34]

[b]Permeability can be calculated by Darcy's Law, $k = \eta LF/(\pi r^2 \triangle P)$, where η is the viscosity, L is the column length, F is the solvent flow rate, r is the radius of trap column, and $\triangle P$ is the column back pressure

[c]The packed length of particulate trap columns was kept at 1 cm due to the pressure limit at high flow rate

porogenic solvent. Monolithic column used in proteome analysis should be stable enough in different separation buffers and not exhibit extra swelling or shrinking, which always influences the permeability of monolith. Therefore, we investigated the permeability of the phosphate monolithic column as well as the particulate SCX and C18 columns packed with 5 μm diameter materials under the solutions of ACN and water, and the obtained results were shown in Table 2.3.

Obviously, the permeability of phosphate monolithic column is 15 times higher than that of particles packed columns under flowing with acetonitrile containing 0.1 % FA and about 10 times higher with 0.1 % FA aqueous solution. It is reported that sulfonate monolith swells in more polar solvents and shrinks in less polar solvent, similar phenomena were also observed in our experiments [34]. However, the permeability decreased only 44 % when the flushing solution was changed from acetonitrile to water for the phosphate monolithic column and relatively low back pressure ~1,300 psi was observed under the high flow rate of water at 40 μL/min. Though slightly swelling in aqueous buffer and shrinking in organic buffer will occur, no detachment of the monolith from the capillary wall was observed at any conditions applied in our experiments. The phosphate monolith was stable enough for continuous usage of 3–4 weeks with a constant back pressure.

As described above, sample loading capacity is one of the most important properties of an ion-exchange chromatography column, which determines the resolution, column capacity, and gradient elution strength. Therefore, a synthetic peptide Leu-Trp-Met-Arg-Phe-NH_2.HCOOH (Mw: 798.42), which bears a +2 charge at pH 2–3, was used to determine the dynamic binding capacity of phosphate SCX monolithic column (150 μm i.d., 7 cm). A sharp increase in the baseline could be observed when the column was saturated (Fig. 2.8). After the phosphate monolithic column was saturated with the peptides, the binding peptides were flushed with 120 μL 1,000 mM NH_4AC buffer by syringe manually. Then, it was reconnected into the system and the capacity was measured again. The binding capacity was measured three times for each of the trap column. Void time estimated by flushing benzoic acid in the same column was subtracted from the total time consumed for saturating. And finally, the average time for saturation was 17.29 min in three measurements (RSD, 1.6 %) for the phosphate monolithic column. As the flow rate was 20 μL/min, the dynamic binding capacity of the phosphate monolithic column was 140 mg/mL, corresponding to 175 μequiv/mL. Binding capacity of particulate SCX column with 150 μm i.d. × 2 cm was also measured following the same procedures. The average time for saturating the particulate SCX column was 4.41 min in three times measurements (RSD, 13.6 %), and the dynamic binding capacity was 125 mg/mL, corresponding to 156 μequiv/mL. However, the packed length of the particulate column was not further increased due to the limitation of the system pressure.

Therefore, the phosphate monolith in unit volume has larger sample loading capacity than the Polysulfoethyl A. Figure 2.8 shows the frontal analysis curves of the synthetic peptide on the SCX monolithic and particulate columns. The frontal analysis curve of phosphate monolithic column increases more sharply,

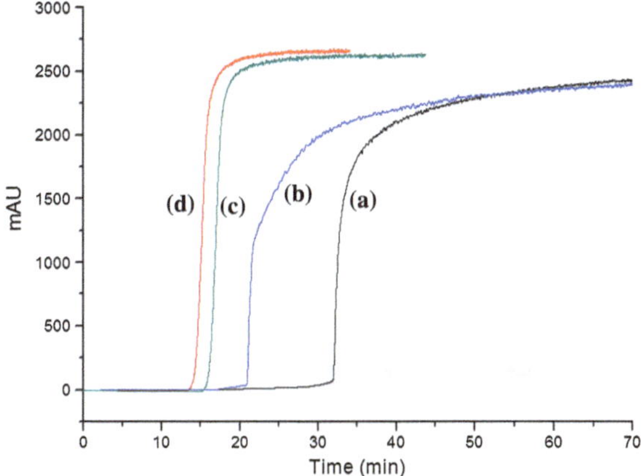

Fig. 2.8 Dynamic adsorption curves of the synthetic peptide by frontal analysis with 7-cm-long phosphate monolithic column (**a**) and 2-cm-long packed polysulfoethyl A column (**b**). The void time was estimated by flushing 50 mM benzonic acid for packed SCX column (**c**) and phosphate monolithic column (**d**), the UV adsorption wavelength was set at 280 nm (Reprinted with permission from Ref. [38]. Copyright 2007 American Chemical Society)

which indicates faster kinetic adsorption of the peptide on the phosphate monolith than on the Polysulfoethyl A particles due to the porous structure of phosphate monolith. Further, the phosphate monolith specifically exhibits SCX mechanism because only one plateau was observed in the frontal analysis curve.

2.2.4 Application of Phosphate SCX Monolithic Column in Automated Sample Injection and SCX-RP 2D LC-MS/MS Analyses

The 150 μm i.d. × 7 cm phosphate SCX monolithic column was used as trap column for fast sample injection followed with RP gradient LC separation by using 75 μm i.d. × 12 cm C18 packed column (Fig. 2.9). First, tryptic digest of 1 μg yeast protein extract was automatically loaded onto the SCX monolithic column at 2 μL/min. Then, the analytes onto the SCX monolithic column were transferred to C18 separation column by 10 min elution with 1,000 mM NH_4AC (pH 2.7) as described in Experimental Section and nanoflow LC-MS/MS was performed (Fig. 2.10). It could be seen that the efficient LC separation time window was about 90 min, which was consistent with the time range from 10 to 35 % ACN. Therefore, little sample loss was occurred during the processes of sample injection. By examining the elution profile of a moderate intensity peptide of K.ATAGDTHLGGEDFDNR.L (m/z = 838, +2) extracted from the separation chromatograms, peak capacity of 359 could be calculated with ~90-min separation window. After database searching and controlling FDR<1 %, with the spectra

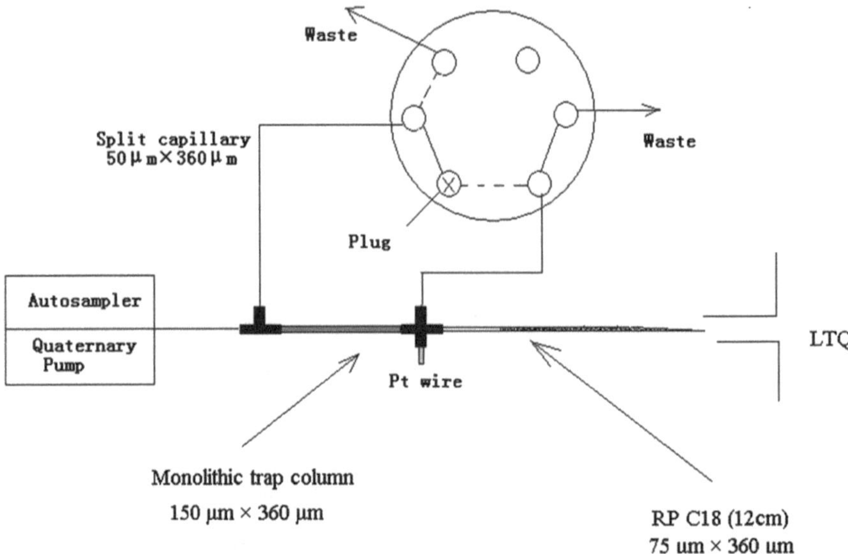

Fig. 2.9 Schematic diagram of the SCX monolithic column online sample injection system. *Solid line* loading sample onto trap column; *dashed line* eluting sample to analytical column and gradient RPLC analysis (Reprinted with permission from Ref. [38]. Copyright 2007 American Chemical Society)

filtered with the above criteria, 1,578 unique peptides were positively identified from 460 distinct proteins of yeast proteome. These results were much better than that obtained by 2 cm packed SCX trap column as described in Sect. 2.2.2. Therefore, increasing the sample loading capacity of trap column is important to improve the performance of proteome analyses.

As the SCX monolithic column has more than 10 times higher permeability than packed SCX column, the sample loading flow rate ranged from 2 to 40 μL/min were invested (Fig. 2.11). Obviously, the flow rate of sample injection has little influence on the numbers of identified peptides and proteins even if the sample injection flow rate increases 20 times. Therefore, sample injection with ultra-high flow rate is feasible, which could improve the throughput of proteome analyses significantly. Then, six times replicate analyses were performed for 1 μg yeast protein and the LC retention time of 10 randomly selected peptides were investigated (Table 2.4). The RSDs for all the retention time of the 10 peptides were almost less than 1 %, which demonstrated the high reproducibility of the automated sample injection system using SCX monolithic trap column in LC separation. Further, 1,556 unique peptides (RSD = 2.0 %) corresponding to 452 distinct proteins (RSD = 0.7 %) were averagely identified from the six replicate analyses, which also demonstrated the good reproducibility of SCX monolithic column system in proteome analyses.

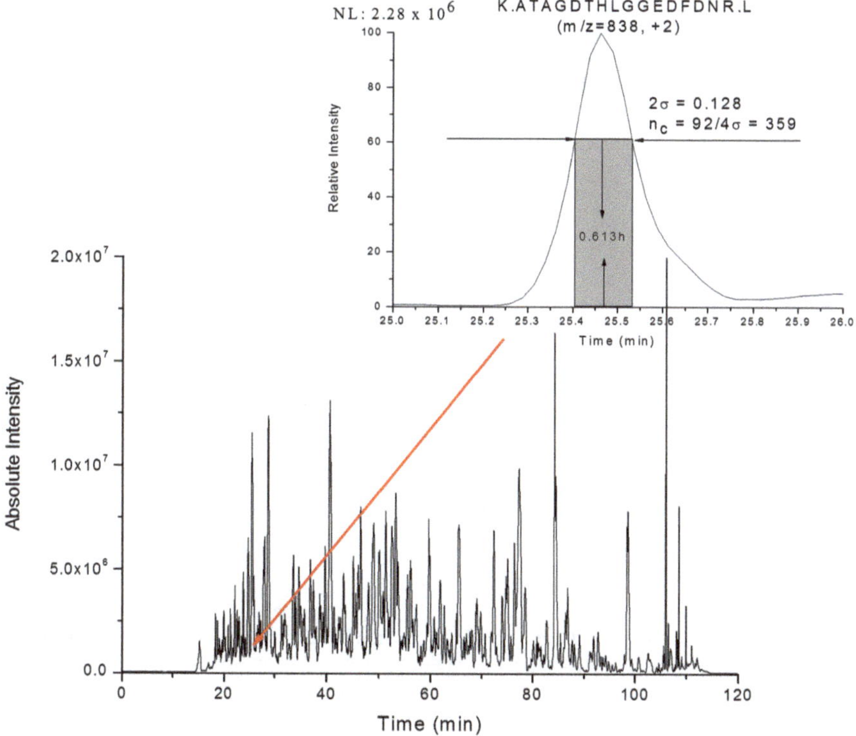

Fig. 2.10 Base peak chromatograms for nano-RPLC-MS/MS analysis of 1 μg yeast proteins tryptic digest on the automated sample injection system using phosphate monolithic trap column (Reprinted with permission from Ref. [38]. Copyright 2007 American Chemical Society)

One-dimensional RP LC separation is usually not enough for comprehensive proteome analysis of complex protein sample as described above. Therefore, SCX-RP 2D LC-MS/MS analysis was also performed for this SCX monolithic column system. After 20 μg tryptic digest of yeast protein extract was automated loaded onto the SCX monolithic column, 17 cycles of salt steps were applied to fractionate the peptides mixture to the C18 separation column, and a nanoflow RP LC MS/MS analysis with 90 min binary RP separation gradient from 10 to 35 % ACN was performed after each fractionation as described in Experimental Section (Fig. 2.12). After database searching and controlling the FDR < 1 % ($\Delta C_n = 0.39$), 5,608 unique peptides (totally 54,780 peptides) corresponding to 1,522 distinct protein groups were successfully characterized. Comparing to the one-dimensional RPLC separation, the proteome coverage was increased 230 %, which is also much better than the results obtained by 2-cm-long SCX packed trap column as described in Sect. 2.2.2. Therefore, the phosphate SCX monolithic column exhibited superior performance in automated sample injection and online 2D

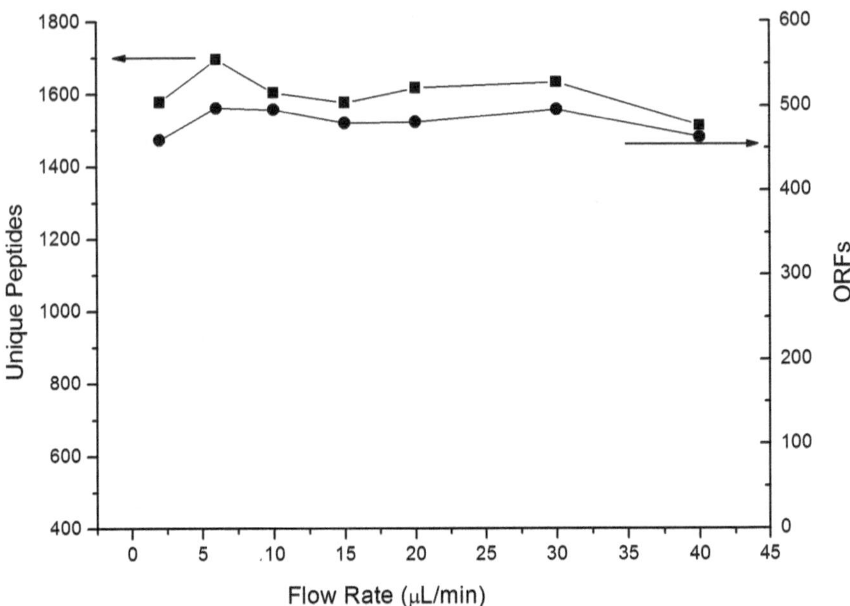

Fig. 2.11 Effect of sample loading flow rate on the number of unique peptides and distinct proteins identified by the automated sample injection system using phosphate monolithic trap column (Reprinted with permission from Ref. [38]. Copyright 2007 American Chemical Society)

Table 2.4 Run-to-run reproducibility of six consecutive nano-RPLC-MS/MS analyses for 1 μg yeast protein tryptic digest by using phosphate monolithic trap column with the automated sample injection*

Mass	Retention time							RSD (%)
	Run 1	Run 2	Run 3	Run 4	Run 5	Run 6	Ave.	
1,017.57	17.84	18.40	18.28	18.41	18.53	18.60	18.34	1.4
883.52	29.20	29.72	29.38	29.22	29.59	29.46	29.43	0.6
1,416.72	35.75	36.29	36.38	36.10	36.42	36.44	36.23	0.7
925.61	43.06	43.70	43.85	43.54	43.87	44.01	43.67	0.7
1,752.79	48.32	49.00	49.01	48.71	49.26	49.18	48.91	0.6
1,351.75	56.22	56.92	56.70	56.67	56.88	56.99	56.73	0.4
2,259.23	68.55	69.19	69.17	68.92	69.00	69.02	68.97	0.3
2,388.20	76.51	76.57	76.69	76.30	76.66	76.66	76.56	0.2
2,831.45	83.39	83.56	83.65	83.59	83.76	83.70	83.61	0.1
1,821.92	88.40	89.35	89.08	89.13	89.36	89.32	89.11	0.4

Reprinted with permission from Ref. [38]. Copyright 2007 American Chemical Society
*Peptides are randomly selected across the elution, and separation conditions are the same as for Fig. 2.10

SCX-RP nanoflow LC-MS/MS for proteome analysis. This can be attributed to its high sample loading capacity and pure SCX retention mechanism, which greatly improve the resolution of SCX fractionation in 2D LC separation.

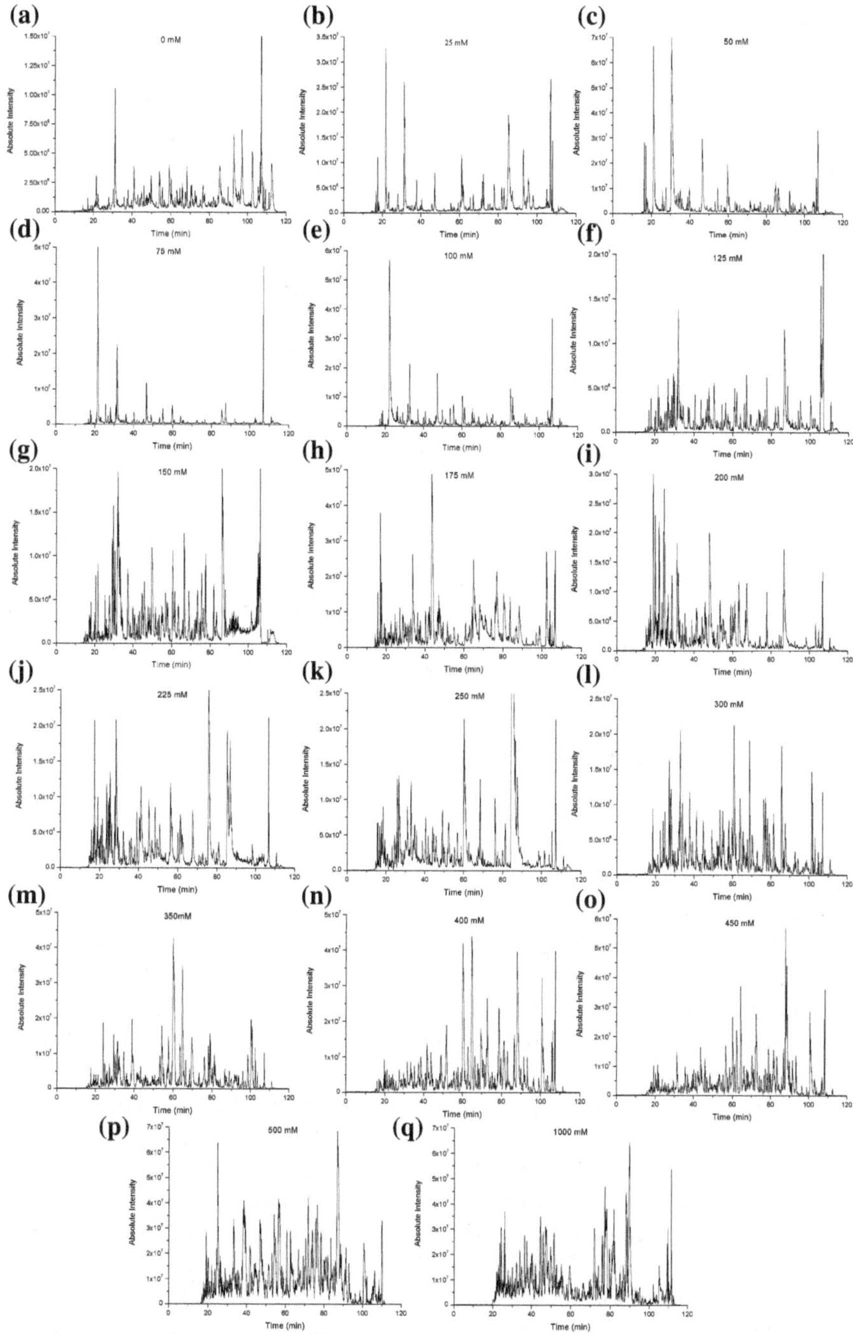

Fig. 2.12 Chromatograms of a 17-cycle online multidimensional analysis of tryptic digest of 20 μg yeast proteins (Reprinted with permission from Ref. [38]. Copyright 2007 American Chemical Society)

2.3 Conclusion

This chapter focuses on the automated sample injection and online multidimensional LC separation in proteome analyses. First, we systematically studied the influence of dead volume between trap and LC separation column on the performance of LC separation and proteome analyses for SCX, C18, and C8 trap column systems. The tolerance of trap column to dead volume is SCX $\gg$ C8 > C18, and SCX trap column system also exhibited the best performance in proteome analyses. Therefore, SCX trap column is more suitable for automated sample injection in nanoflow LC-MS/MS system. Then, an automated sample-clean system was established by SCX trap column and 10-port nanoswitching valve, and superior performance was obtained for proteome analysis of uncleaned protein digest due to the low protein sample loss in online purification. Finally, a phosphate SCX monolithic column was prepared for automated sample injection and online fractionation of the peptides mixture. Comparing with packed SCX column, 10 times higher permeability was exhibited, and sample injection flow rate as high as 40 μL/min could be realized. Furthermore, high sample loading capacity was also achieved for the SCX monolithic column, which significantly improved the resolution in online fractionation and also for the proteome coverage in SCX-RP 2D LC-MS/MS analyses.

2.4 Experimental Section

2.4.1 Materials

Daisogel ODS-AQ (5 μm, 12 nm pore) was purchased from DAISO Chemical CO., Ltd. (Osaka, Japan), Magic C18AQ (5 μm, 20 nm pore) was purchased from Michrom BioResources (Auburn, CA, USA), and Polysulfoethyl Aspartamide (5 μm, 20 nm pore) was a gift from PolyLC Inc. (Columbia, MD, USA). PEEK tubing, sleeves, microtee, microcross, and mini microfilter assembly (with a filter capsule, porosity 2 μm) were obtained from Upchurch Scientific (Oak Harbor, WA, USA). Fused silica capillaries with 75 and 150 μm i.d. were purchased from Polymicro Technologies (Phoenix, AZ, USA), and 50 μm i.d. from Yongnian Optical Fiber Factory (Hebei, China). All the water used in experiments was purified using a Mill-Q system from Millipore Company (Bedford, MA, USA). Dithiothreitol (DTT), iodoacetamide were both purchased from Sino-American Biotechnology Corporation (Beijing, China). Ethylene glycol methacrylate phosphate (EGMP), bis-acrylamide, dimethylsulfoxide, dodecanol, N, N'-dimethylformamide, and γ-methacryloxypropyltrimethoxysilane (γ-MAPS) were obtained from Sigma (St. Louis, MO, USA). Azobisisobutyronitrile (AIBN) was obtained from Shanghai Fourth Reagent Plant (Shanghai, China). Trypsin was obtained from Promega (Madison, WI, USA). Synthetic peptide was purchased from Serva (Heidelberg, Germany), and Tris was from Amersco (Solon, Ohio, USA). Formic acid was obtained from Fluka (Buches, Germany). Acetonitrile (ACN, HPLC grade) was from Merck (Darmstadt, Germany).

2.4.2 Sample Preparation

The yeast protein extract was prepared in a denaturing buffer containing 50 mM Tris/HCl (pH 8.1) and 8 M urea. The protein concentration was determined by BCA assay. The protein sample was reduced by DTT at 37 °C for 2 h and alkylated by iodoacetamide in dark at room temperature for 40 min. Then the solution was diluted to 1 M urea with 50 mM Tris/HCl (pH 8.1). Finally, trypsin was added with weight ratio of trypsin to protein at 1/25 and incubated at 37 °C overnight. Then, the tryptic digest was purified with a homemade C18 solid phase cartridge and exchanged into 0.1 % FA water solution. Finally, the samples were stored at −20 °C before usage.

2.4.3 Preparation of Particles Packed Columns

Particulate columns were packed using a homemade pneumatic pressure cell at constant nitrogen gas pressure of about 580 psi with a slurry packing method. For the preparation of C18 separation column, one end of a 75 μm-i.d. fused silica capillary was manually pulled to a fine point of ~5 μm with a flame torch at first, and then 12-cm-long C18 particles (5 μm, 12 nm pore) were packed. After a mini microfilter assembly (with a filter capsule, porosity 2 μm) was placed on one end of a 150 μm-i.d. capillary, 1- or 2-cm-long SCX or C18 material (5 μm, 20 nm pore) was packed into the column to prepare the particulate SCX or C18 trap column. After the column was packed with appropriate length, it was connected to a LC pump and equilibrated with 0.1 % FA water solution at ~1,500 psi for at least 30 min before usage. The 2-cm-long particulate SCX column was used to measure the dynamic binding capacity of Polysulfoethyl A and 1-cm-long particulate SCX and C18 trap columns were used in permeability measurement. 2-cm-long particulate trap columns were also used for sample injection in proteome analysis.

2.4.4 Preparation of Phosphate Monolithic Capillary Column

Prior to the polymerization, the capillary was pretreated with γ-MAPS [37]. The reaction mixture consisting of EGMP (80 μL, ~100 mg), bis-acrylamide (60 mg), dimethylsulfoxide (270 μL), dodecanol (200 μL), N, N′-dimethylformamide (50 μL), and AIBN (2 mg) was sonicated for 20 min to obtain a homogeneous solution and then purged with nitrogen for 10 min. After the pretreated capillary was completely filled with the mixture, it was sealed at both ends with rubber stoppers. The sealed capillary was submerged into a water bath and allowed to react for 12 h at 60 °C. The resultant monolithic capillary column was washed with methanol for 2 h using an HPLC pump to remove unreacted monomers and porogens. Then, this phosphate monolithic capillary column could be directly used as the SCX trap column for proteome analysis without any other pretreatments.

Scanning electron microscopy (SEM) images of the monolithic column were obtained using a JEOL JSM-5,600 scanning electron microscope (JEOL Company, Japan).

2.4.5 Binding Capacity Measurement of the Phosphate Monolithic Capillary Column

For measurement of the dynamic binding capacity, a synthetic peptide Leu-Trp-Met-Arg-Phe-NH2.HCOOH (Mw: 798.42) was used to saturate the phosphate monolithic column (150 μm i.d., 7 cm) by frontal analysis. Briefly, 10 mg synthetic peptide was dissolved into 20 mL buffer containing 0.1 % FA and 10 % acetonitrile (pH 2–3), which was pumped through the phosphate monolith column at a constant flow rate of 20 μL/min by a LC-10ADvp pump (Shimadzu). The solution from the monolithic column flowed into a 500 nL 1,100-DAD micro cell UV detector (Agilent). A 50-mM benzonic acid water solution containing 0.1 % FA and 10 % acetonitrile was used to estimate the void time of this system at the same flow rate because benzonic acid with considerable UV absorption is neutral in acidic solution.

To provide comparable data, the dynamic binding capacity of a particulate SCX column with 150 μm i.d. × 2 cm was also measured following the same procedures.

2.4.6 Sample Injection

The configuration for offline sample injection system by using C18, C8, or SCX trap columns was shown in Fig. 2.1. This system was used to systematically study the influence of dead volume between trap and analytical columns on the separation performance and proteomic coverage. During sample loading, the 6-port/two-position valve was switched to close the split flow, and 1 μg SPE cleaned yeast protein digest (20 μL) was injected by Surveyor autosampler using the no-waste injection mode at a flow rate of 2 μL/min as shown in Fig. 2.1a. The mobile phase used was 0.1 % formic acid. After sample injection, the trap column and C18 separation column were manually connected by a ZDV (zero dead volume) union. Then the valve was switched to start the split flow in the front of trap column. In order to investigate the influence of void volume on separation, appropriate void volume was introduced between trap and separation columns by using 150-μm-i.d. void capillary as shown in Fig. 2.1b. And 5.7-, 16.9-, and 28.3-cm-long 150-μm-i.d. void capillary were used to generate 1, 3, and 5 μL dead volume, respectively.

The configuration for automated online sample injection of uncleaned protein sample using SCX or C18 trap column was shown in Fig. 2.5. The trap column was connected to a C18 separation column by a nanoflow switching valve and a micro-tee. During sample injection, the switching valve was switched to close the splitting flow and the flow through from the trap column was directed to waste. Yeast protein tryptic digest without SPE pretreatment (20 μL) was directly injected (Fig. 2.5 solid line mode) including 10 min for sample loading and 20 min for equilibration. After sample loading, the switching valve was switched to activate the splitting flow in the front of trap column and the flow through trap column was switched to the separation column at a rate ~200 nL/min (Fig. 2.5 dashed line mode).

For manual injection, an open capillary filled with sample was connected between the microcross and the analytical column [39]. The peptides in the open capillary was flushed onto the analytical column by the 0.1 % FA at a flow rate of ~200 nL/min (after splitting) and retained on the analytical column front. After sample loading, the open capillary was removed, and the separation column was directly connected to the microcross.

The configuration for automated sample injection using phosphate monolithic trap column was shown in Fig 2.9. During sample injection, the switching valve was switched to close the splitting flow and the flow through from the trap column was switched to waste (solid line mode in Fig. 2.9). We adjusted the time of sample injection at different flow rates to ensure the volume that flow though the trap column was three sample volumes for system equilibrium and removing contaminants. The sample solutions were loaded at flow rate of 2, 6, 10, 15, 20, 30, 40 μL/min by setting loading times at 30, 10, 6, 4, 3, 2, 1.5 min, respectively, when 20 μL sample was injected. After sample loading, the valve was switched to the dashed line mode and the gradient separation was automatically started.

2.4.7 One-Dimensional LC Separation

The three buffer solutions used for the quaternary pump were 0.1 % FA water solution (buffer A), ACN with 0.1 % FA (buffer B), 1,000 mM NH_4Ac at pH 3 (buffer C). The flow rate after splitting was adjusted to separation optimal ~180 nL/min. The binary gradient with buffer A and buffer B for RP separation was developed from 0 to 10 % buffer B for 2 min, from 10 to 35 % for 90 min, and from 35 to 80 % for 5 min. After flushing 80 % buffer B for 10 min, the separation system was equilibrated by buffer A for 13 min. In one-dimensional separation, the peptides retained on trap column were all eluted onto separation column by flushing with buffer C (containing 1,000 mM NH_4Ac at pH 3) for 10 min. After the system was re-equilibrated with buffer A for 10 min, the binary separation gradient described above was started.

2.4.8 Multidimensional LC Separation

20 μg tryptic digest of yeast protein was loaded automatically onto the particulate or monolithic SCX trap column. Then, a series of stepwise elution with salt concentrations of 0, 25, 50, 75, 100, 125, 150, 175, 200, 225, 250, 300, 350, 400, 450, 500, and 1000 mM NH_4Ac was used to gradually elute peptides from phosphate monolithic column onto the C18 analytical column (for particulate SCX trap column, 75, 125, 175, and 225 mM were not performed). Each salt step lasts 5 min except last one for 10 min. After whole system was re-equilibrated for 10 min with buffer A, the binary gradient elution described in Sect. 2.4.7 was applied to separate peptides prior to MS detection in each cycle.

2.4.9 Mass Spectrometric Analysis

The temperature of the ion transfer capillary was set at 200 °C. The spray voltage was set at 1.82 kV and the normalized collision energy was set at 35.0 %. One microscan was set for each MS and MS/MS scan. All MS and MS/MS spectra were acquired in the data-dependent mode. The mass spectrometer was set that one full MS scan was followed by six MS/MS scans on the six most intense ions. The dynamic exclusion function was set as follows: repeat count two, repeat duration 30 s, and exclusion duration 90 s. System control and data collection were done by Xcalibur software version 1.4 (Thermo).

2.4.10 Data Analysis

The acquired MS/MS spectra were searched on the database using the Turbo SEQUEST in the BioWorks 3.2 software suite (Thermo). The yeast database was downloaded from a website (ftp://genome-ftp.stanford.edu/yeast/data_download/sequence/genomic_sequence/orf_protein/orf_trans.fasta.gz). Reversed sequences were appended to the database for the evaluation of false positive rate. Cysteine residues were searched as static modification of 57.0215 Da, and methionine residues as variable modification of +15.9949 Da. Peptides were searched using fully tryptic cleavage constraints and up to two internal cleavages sites were allowed for tryptic digestion. The mass tolerances were 2 Da for parent masses and 1 Da for fragment masses. The peptides were considered as positive identification if the Xcorr were higher than 1.9 for singly charged peptide, 2.2 for doubly charged peptide, and 3.75 for triply charged peptides. ΔCn cutoff value was set to control the false positive rate of peptides identification <1 %, determined by the calculation based on the reversed database 12.

References

1. Link AJ, Eng J, Schieltz DM, Carmack E, Mize GJ, Morris DR, Garvik BM, Yates JR (1999) Direct analysis of protein complexes using mass spectrometry. Nat Biotechnol 17:676–682
2. Wolters DA, Washburn MP, Yates JR (2001) An automated multidimensional protein identification technology for shotgun proteomics. Anal Chem 73:5683–5690
3. Jiang X, Feng S, Tian R, Han G, Ye M, Zou H (2007) Automation of nanoflow liquid chromatography-tandem mass spectrometry for proteome analysis by using a strong cation exchange trap column. Proteomics 7:528–539
4. Ishihama Y (2005) Proteomic LC-MS systems using nanoscale liquid chromatography with tandem mass spectrometry. J Chromatogr A 1067:73–83
5. van der Heeft E, ten Hove GJ, Herberts CA, Meiring HD, van Els CA, de Jong AP (1998) A microcapillary column switching HPLC-electrospray ionization MS system for the direct identification of peptides presented by major histocompatibility complex class I molecules. Anal Chem 70:3742–3751
6. Shen Y, Tolić N, Masselon C, Paša-Tolić L, Camp DG, Hixson KK, Zhao R, Anderson GA, Smith RD (2003) Ultrasensitive proteomics using high-efficiency on-line micro-SPE-nanoLC-nanoESI MS and MS/MS. Anal Chem 76:144–154
7. Shen Y, Zhang R, Moore RJ, Kim J, Metz TO, Hixson KK, Zhao R, Livesay EA, Udseth HR, Smith RD (2005) Automated 20 kpsi RPLC-MS and MS/MS with chromatographic peak capacities of 1000–1500 and capabilities in proteomics and metabolomics. Anal Chem 77:3090–3100
8. Licklider LJ, Thoreen CC, Peng J, Gygi SP (2002) Automation of nanoscale microcapillary liquid chromatography-tandem mass spectrometry with a vented column. Anal Chem 74:3076–3083
9. Meiring HD, Van Der Heeft E, Ten Hove GJ, De Jong APJM (2002) Nanoscale LC-MS: technical design and applications to peptide and protein analysis. J Sep Sci 25:557–568
10. Yi EC, Lee H, Aebersold R, Goodlett DR (2003) A microcapillary trap cartridge-microcapillary high-performance liquid chromatography electrospray ionization emitter device capable of peptide tandem mass spectrometry at the attomole level on an ion trap mass spectrometer with automated routine operation. Rapid Commun Mass Spectrom 17:2093–2098
11. Kang D, Nam H, Kim YS, Moon MH (2005) Dual-purpose sample trap for on-line strong cation-exchange chromatography/reversed-phase liquid chromatography/tandem mass spectrometry for shotgun proteomics: application to the human jurkat T-cell proteome. J Chromatogr A 1070:193–200
12. Wu L, Han DK (2006) Overcoming the dynamic range problem in mass spectrometry-based shotgun proteomics. Expert Rev Proteomics 3:611–619
13. Shen Y, Moore RJ, Zhao R, Blonder J, Auberry DL, Masselon C, Paša-Tolić L, Hixson KK, Auberry KJ, Smith RD (2003) High-efficiency on-line solid-phase extraction coupling to 15–150 μm-i.d. Column liquid chromatography for proteomic analysis. Anal Chem 75:3596–3605
14. Shen Y, Zhao R, Berger SJ, Anderson GA, Rodriguez N, Smith RD (2002) High-efficiency nanoscale liquid chromatography coupled on-line with mass spectrometry using nanoelectrospray ionization for proteomics. Anal Chem 74:4235–4249

15. Vollmer M, Hörth P, Nägele E (2004) Optimization of two-dimensional off-line LC/MS separations to improve resolution of complex proteomic samples. Anal Chem 76:5180–5185
16. Peng J, Elias JE, Thoreen CC, Licklider LJ, Gygi SP (2003) Evaluation of multidimensional chromatography coupled with tandem mass spectrometry (LC/LC-MS/MS) for large-scale protein analysis: the yeast proteome. J Proteome Res 2:43–50
17. Wagner K, Miliotis T, Marko-Varga G, Bischoff R, Unger KK (2002) An automated on-line multidimensional HPLC system for protein and peptide mapping with integrated sample preparation. Anal Chem 74:809–820
18. Xiang R, Shi Y, Dillon DA, Negin B, Horváth C, Wilkins JA (2004) 2D LC/MS analysis of membrane proteins from breast cancer cell lines MCF7 and BT474. J Proteome Res 3:1278–1283
19. Opiteck GJ, Jorgenson JW, Anderegg RJ (1997) Two-dimensional SEC/RPLC coupled to mass spectrometry for the analysis of peptides. Anal Chem 69:2283–2291
20. Lohaus C, Nolte A, Blüggel M, Scheer C, Klose J, Gobom J, Schüler A, Wiebringhaus T, Meyer HE, Marcus K (2006) Multidimensional chromatography: a powerful tool for the analysis of membrane proteins in mouse brain. J Proteome Res 6:105–113
21. Gilar M, Daly AE, Kele M, Neue UD, Gebler JC (2004) Implications of column peak capacity on the separation of complex peptide mixtures in single- and two-dimensional high-performance liquid chromatography. J Chromatogr A 1061:183–192
22. Washburn MP, Wolters D, Yates JR (2001) Large-scale analysis of the yeast proteome by multidimensional protein identification technology. Nat Biotechnol 19:242–247
23. Zou H, Huang X, Ye M, Luo Q (2002) Monolithic stationary phases for liquid chromatography and capillary electrochromatography. J Chromatogr A 954:5–32
24. Svec F (2004) Preparation and HPLC applications of rigid macroporous organic polymer monoliths. J Sep Sci 27:747–766
25. Svec F (2004) Organic polymer monoliths as stationary phases for capillary HPLC. J Sep Sci 27:1419–1430
26. Xie C, Ye M, Jiang X, Jin W, Zou H (2006) Octadecylated silica monolith capillary column with integrated nanoelectrospray ionization emitter for highly efficient proteome analysis. Mol Cell Proteomics 5:454–461
27. Wu R, Zou H, Ye M, Lei Z, Ni J (2001) Separation of basic, acidic and neutral compounds by capillary electrochromatography using uncharged monolithic capillary columns modified with anionic and cationic surfactants. Electrophoresis 22:544–551
28. Liu Z, Wu R, Zou H (2002) Recent progress in adsorbed stationary phases for capillary electrochromatography. Electrophoresis 23:3954–3972
29. Viklund C, Svec F, Frechet JM, Irgum K (1997) Fast ion-exchange HPLC of proteins using porous poly(glycidyl methacrylate-co-ethylene dimethacrylate) monoliths grafted with poly(2-acrylamido-2-methyl-1-propanesulfonic acid). Biotechnol Prog 13:597–600
30. Ueki Y, Umemura T, Li J, Odake T, Tsunoda K (2004) Preparation and application of methacrylate-based cation-exchange monolithic columns for capillary ion chromatography. Anal Chem 76:7007–7012
31. Peters EC, Petro M, Svec F, Fréchet JMJ (1997) Molded rigid polymer monoliths as separation media for capillary electrochromatography. Anal Chem 69:3646–3649
32. Peters EC, Petro M, Svec F, Fréchet JMJ (1998) Molded rigid polymer monoliths as separation media for capillary electrochromatography. 1. Fine control of porous properties and surface chemistry. Anal Chem 70:2288–2295
33. Wu R, Zou H, Fu H, Jin W, Ye M (2002) Separation of peptides on mixed mode of reversed-phase and ion-exchange capillary electrochromatography with a monolithic column. Electrophoresis 23:1239–1245
34. Gu B, Chen Z, Thulin CD, Lee ML (2006) Efficient polymer monolith for strong cation-exchange capillary liquid chromatography of peptides. Anal Chem 78:3509–3518

35. Wang F, Jiang X, Feng S, Tian R, Han G, Liu H, Ye M, Zou H (2007) Automated injection of uncleaned samples using a ten-port switching valve and a strong cation-exchange trap column for proteome analysis. J Chromatogr A 1171:56–62
36. Fu H, Xie C, Dong J, Huang X, Zou H (2004) Monolithic column with zwitterionic stationary phase for capillary electrochromatography. Anal Chem 76:4866–4874
37. Dong J, Zhou H, Wu R, Ye M, Zou H (2007) Specific capture of phosphopeptides by Zr^{4+}-modified monolithic capillary column. J Sep Sci 30:2917–2923
38. Wang F, Dong J, Jiang X, Ye M, Zou H (2007) Capillary trap column with strong cation-exchange monolith for automated shotgun proteome analysis. Anal Chem 79:6599–6606
39. Feng S, Ye M, Jiang X, Jin W, Zou H (2006) Coupling the immobilized trypsin microreactor of monolithic capillary with μRPLC−MS/MS for shotgun proteome analysis. J Proteome Res 5:422–428

Chapter 3
Development of Polymer-Based Hydrophobic Monolithic Columns and Their Applications in Proteome Analysis

3.1 Introduction

As described in Chap. 2, the LC separation capability is one of the most important factors to achieve high proteome coverage in shotgun proteome analysis. In order to improve the separation efficiency, long columns packed with smaller particles are commonly applied [1–3]. Though the operating pressure of the ultra-high performance liquid chromatography (UPLC) instrument is often over 10,000 psi, the packed length of separation column is still limited, especially when capillary columns with smaller inner diameters (<50 μm) are utilized in nanoflow liquid chromatography coupled with tandem mass spectrometry (LC-MS/MS) analysis. Further, the column packing is difficult and time-consuming when column with smaller dimension and smaller packing material (<5 μm) is utilized. Monolithic columns have higher permeability and faster mass transferring rate than columns packed with particle materials due to its porous structure, which makes them good alternatives to packed columns in chromatography separation. Especially in shotgun proteome analysis, where capillary column (usually 10–100 μm i.d.) is adopted to increase the sensitivity as well as extremely complex protein samples such as serum is inevitably confronted, monolithic capillary columns are feasibly explored to improve the LC separation performance.

Although silica-based inorganic monolithic capillary column can greatly improve the LC separation performance for tryptic peptides due to its high-uniform porous structure, the processes for the preparation of silica-based monolithic column are difficult and time-consuming [4–8]. Further, post-modification of the surface of silica-based monolith with functional monomers is needed after the preparation of monolithic framework and the unreacted active groups also need to be blocked by additional reaction. In contrast to inorganic monolithic column, polymer-based organic monolithic column is easy to be prepared by just one-step copolymerization of functional monomer and crosslinker, and a number of functional monomer candidates can be utilized to obtain organic monolith with different LC retention mechanisms, such as reversed phase (RP), strong-cation exchange

F. Wang, *Applications of Monolithic Column and Isotope Dimethylation Labeling in Shotgun Proteome Analysis*, Springer Theses, DOI: 10.1007/978-3-642-42008-5_3,

(SCX), and strong anion exchange (SAX) [9–13]. Poly-(styrene-divinylbenzene) (PS-DVB) is one of the most popular functional monomers for the preparation of RP organic monolithic column (both commercially available and homemade), which was widely applied for peptide or protein RPLC separation [14, 15]. By coupling to SCX chromatography, the PS-DVB monolithic capillary column can be applied in offline two-dimensional (2D) separation of proteolytic digest to obtain high proteome coverage [16, 17]. The PS-DVB monolithic capillary column was also used to analyze the phosphopeptides in alkaline eluent (pH 9.2) in conjunction with MS detection in negative ion mode due to its wide operating pH range [18]. A 4.2-m-long PS-DVB porous layer open tubular (PLOT) capillary column with inside diameter of only 10 μm was further developed to increase the detection limit as well as the separation efficiency, and this PLOT column was successfully applied in high-resolution 1- or 2D nanoflow LC-MS/MS analysis of limited amount of protein sample [19, 20].

Methacrylate-based monolithic column is believed to be rigid among all of the polymer-based organic monolithic columns, and it can be easily prepared by direct copolymerization of the methacrylate functional monomer and crosslinker [9–11]. Wu et al. prepared a neutrally hydrophobic monolithic column by in situ copolymerization of lauryl methacrylate (LMA) and ethylene dimethacrylate (EDMA) to form a C12 hydrophobic stationary phase, and it was successfully applied in capillary electrochromatography (CEC) for separation of ionic compounds driven by electrophoretic mobility [21]. We wonder if this hydrophobic C12 monolithic column can be also used for high-efficient RPLC separation of complex protein sample in proteomics. Thus, this chapter investigated the different applications of C12 RP monolithic column in proteomics. We studied the effect of constitute and percentage of porogenic solvent, functional monomer, column length, and RP separation gradient on separation capability and the coverage of yeast proteome by hydrophobic methacrylate-based monolithic column. The homogeneity of the pore structure (both pores and microglobules), which was mainly determined by the porogenic solvent, was crucial to the peak capacity and proteome coverage when similar methacrylate monomer was utilized. The hydrophobic C12 monolithic column was also coupled with the phosphate SCX monolithic column described in Chap. 2 for online 2D LC-MS/MS analysis to further improve the detected proteome coverage. Besides LC separation, we also investigate the other applications of C12 monolithic column in proteomics. As the C12 monolith exhibited superior permeability than packed column, we introduced the hydrophobic porous monolith into the electro-spray ionization (ESI) emitter to reduce the post-column void volume as well as the disruption of LC separation profile. It was observed that both the peak capacity of the LC separation and the proteome quantification accuracy were significantly improved after utilization of the C12 ESI emitter. Finally, to alleviate the difficulty of solid phase material packing, a C12 monolithic frit was fabricated onto one end of the void capillary. And an integrated monolithic ESI emitter was prepared after the packing of LC solid phase material. Therefore, smaller LC solid phase material could be easily packed into the capillary column, which improved the LC separation capability and proteome coverage greatly.

3.2 Experimental Results

3.2.1 LC Separation Performance and Proteome Coverage of the Reversed Phase C12 Monolithic Column

3.2.1.1 Effect of Porogenic Solvent

We investigated the effect of porogenic solvent on the porous structure of C12 monolith and the performance of proteome analysis at first. Three types of C12 monolithic columns were synthesized with the constitute ratio of LMA:EDMA:1-propanol:1,4-butanediol:water at 100:100:170:130:0, 100:100:147:73:0, and 100:100:170:130:20, respectively, and the scanning electron microscopy photographs are shown in Fig. 3.1. The first type of C12 column exhibits the best permeability as shown in Table 3.1, and it can be also seen that macropores as large as about 5 μm are formed into the monolith (Fig. 3.1a). However, the proteome coverage of this C12 column was poor and only 1,047 unique peptides (RSD = 2.7 %, $n = 3$) and 385 proteins (RSD = 5.6 %, $n = 3$) were positively identified by injection of 1 μg tryptic digest of yeast protein and separating with 210-min gradient on 85 cm × 75 μm i.d. column. The extracted ion currents (XICs) of three identified peptides K.ISHVSTGGGASLELLEGK.E, K.YGVTDKISHVSTGGGASLELLEGK.E, and K.DKADVDGFLVGGASLKPEFVDIINSR.N were selectively extracted from the base peak chromatogram (Fig. 3.2a), and it could be seen that the poor proteome coverage was induced by the poor LC separation performance, which was influenced greatly by the inhomogeneity of the overall pore structure (both pores and microglobules) of the C12 monolith [22, 23]. It can be also seen that the average peak width at 0.613 height of the peak was 0.81 min, and the average peak capacity 129 could be obtained (210-min efficient separation time) [5]. Therefore, we decreased the percentage of binary porogenic solvent of 1-propanol and 1,4-butanediol in the polymerization mixture as described above and the second type of C12 monolithic column was synthesized with relative low permeability as shown in Table 3.1. The monolith of this C12 column is more compact and the pore structure is smaller as shown in Fig. 3.1b. When the same test sample and separation gradient were applied, 1,787 unique peptides (RSD = 2.3 %, $n = 3$) and 574 proteins (RSD = 0.1 %, $n = 3$) were positively identified. Therefore, the identified unique peptides and proteins were increased by 70.7 and 49.1 % compared to the first type of C12 monolithic column. XICs of the same three peptides were also extracted and are shown in Fig. 3.2b, which exhibits better LC elution profiles and much higher peak intensity. The average peak width at 0.613 height of the peak was 0.75 min, and average peak capacity 141 was obtained. All of the above two C12 monolithic columns were prepared with binary porogenic solvent of 1-propanol and 1,4-butanediol. Then, a ternary porogenic solvent of 1-propanol, 1,4-butanediol, and water (170:130:20) was further adopted to prepare the third type of C12 monolithic column. Compared to the first C12 monolithic column, the only difference was that

Fig. 3.1 Scanning electron microscopy photographs at magnification of 2,500× for three types of C12 monolithic columns with the constitute ratio of LMA:EDMA:1-propanol:1,4-butanediol:water at **a** 100:00:170:130:0, **b** 100:100:147:73:0, and **c** 100:100:170:130:20, and **d** one type of C18 monolithic column with the constitute of SMA:EDMA:1-propanol:1,4-butanediol at 100:100:133:67 (Reprinted with permission from Ref. [24]. Copyright 2009 Elsevier)

20 μL water was added into the polymerization mixture. It was observed that the C12 monolithic column prepared with porogenic solvent containing 20 μL water had the lowest permeability as shown in Table 3.1 and the overall pore structure was smaller than the first type of C12 monolithic column and more homogeneous than the second type of C12 monolithic column as shown in Fig. 3.1c. The base peak chromatogram of nanoflow LC-MS/MS analysis of 1 μg tryptic digest of yeast protein is shown in Fig. 3.3, and the elution profile is symmetrically distributed among the ~210-min efficient separation time. Finally, 2,081 unique peptides (RSD = 2.4 %, n = 3) and 614 proteins (RSD = 2.0 %, n = 3) were positively identified. The identified unique peptides and proteins were increased 98.8, 59.5 % and 16.4, 7.0 % compared to the first and second C12 monolithic columns. Then, XICs of the three peptides were extracted and shown in Fig. 3.2c, the average

Table 3.1 Permeability of C12 and C18 monolithic columns with the 85 cm × 75 μm i.d. prepared in different porogenic solvents

Column (85 cm × 75 μm i.d.)	Flushing solution	Viscosity, η (cP)[a]	Back pressure, $\triangle P$ (psi)	Flow rate, F (nL/min)	Permeability, k ($\times 10^{-14}$ m^2)[b]
C12-1	Acetonitrile	0.369	326	200	10.6
	Water	0.890	733	200	11.4
C12-2	Acetonitrile	0.369	737	200	4.7
	Water	0.890	1,348	200	6.2
C12-3	Acetonitrile	0.369	1,020	200	3.4
	Water	0.890	1,774	200	4.7
C18	Acetonitrile	0.369	952	200	3.6
	Water	0.890	1,462	200	5.7

[a] Viscosity data were from Ref. [25]
[b] Permeability can be calculated by Darcy's Law [26], $k = \eta LF/(\pi r^2 \triangle P)$, where η is the viscosity, L is the column length, F is the solvent flow rate, r is the radius of the column, and $\triangle P$ is the column backpressure

peak width at 0.613 height of the peak was 0.61 min and the average peak capacity was 173, which further demonstrated that the best LC separation capability was obtained compared to the other two columns. To investigate the column-to-column reproducibility of the C12 monolithic column, three different C12 monolithic columns were fabricated in three different batches by using the ternary porogenic solvent. After nanoflow LC-MS/MS analyses of 1 μg tryptic digest of yeast protein, the retention time of 12 randomly selected peptides were investigated as shown in Table 3.2. Obviously, the RDSs of three C12 monolithic columns in different batches were all less than 5 %, which demonstrated good column-to-column reproducibility was achieved in the preparation of C12 monolithic column.

The overall pore structure of monolithic column is crucial to the chromatographic separation. The first type of C12 monolithic column had the largest pore size (revealed by the largest permeability), which induced the smallest surface area and lowest retentivity and selectivity in separation process. In order to decrease the pore size of the monolith, the percentage of the binary porogenic solvent was decreased from 60 to 52.4 % and the percentage of 1-propanol in the porogenic solvent was increased from 56.7 to 66.7 % as described by Peters et al. [22]. Though the permeability of C12 monolithic column was decreased, the proteome coverage increased greatly as expected. When the methacrylate-based monolithic column was applied into CEC, water was always added into the polymerization mixture to dissolve the solid 2-acrylamido-2-methyl-1-propane-sulfonic acid (AMPS), which was a charged monomer added to induce electroosmotic flow [23]. In the pressure driving chromatography, AMPS and water were not usually utilized for the preparation of methacrylate-based monolithic columns. However, when 20 μL water was added into polymerization mixture to prepare C12 monolithic column, the permeability of the column decreased and the proteome

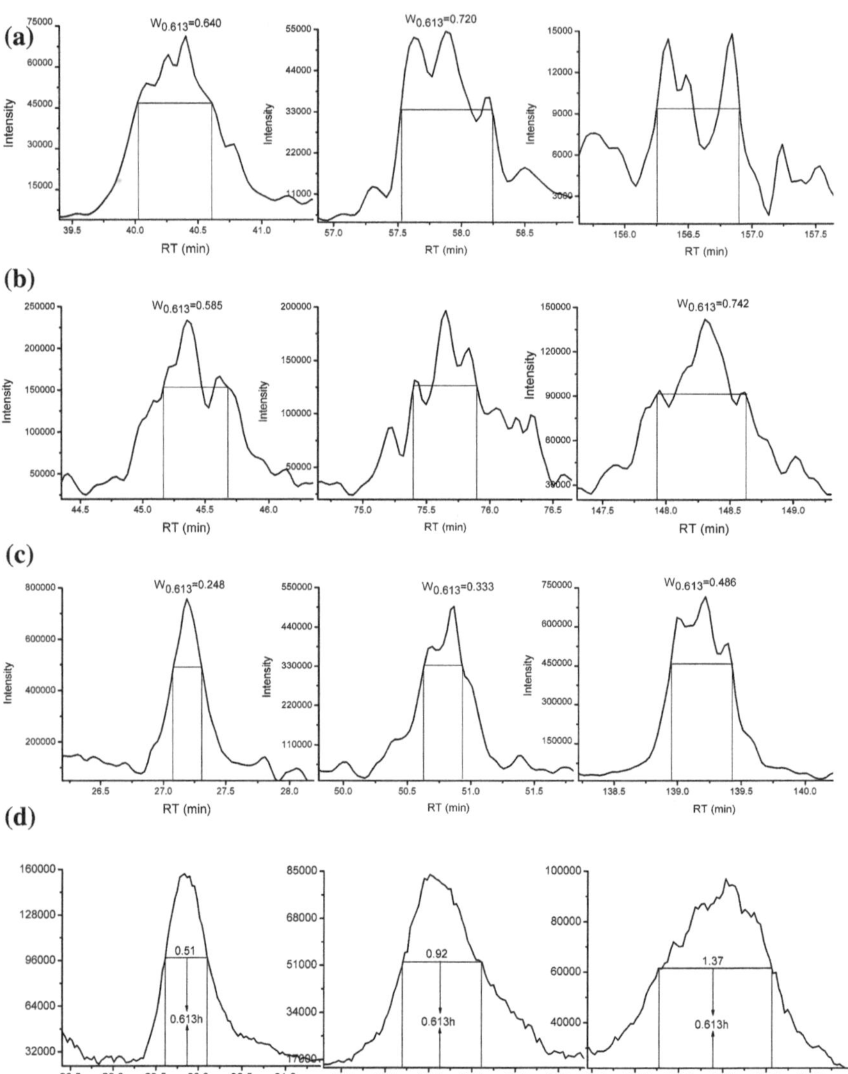

Fig. 3.2 Single ion currents of three identified peptides K.ISHVSTGGGASLELLEGK.E, K.YGVTDKISHVSTGGGASLELLEGK.E, and K.DKADVDGFLVGGASLKPEFVDIINSR.N for **a** the first type of C12 monolithic column, **b** the second type of C12 monolithic column, **c** the third type of C12 monolithic column, and **d** C18 monolithic column ($W_{0.613}$, peak width at 0.613 height of the peak, each $W_{0.613}$ was an average value of three replicated analysis) (Reprinted with permission from Ref. [24]. Copyright 2009 Elsevier)

coverage increased evidently. This implied that the overall pore structure became more homogeneous and compact in despite of the increase of the percentage of porogenic solvent. Therefore, not only the percentage of porogenic solvent within

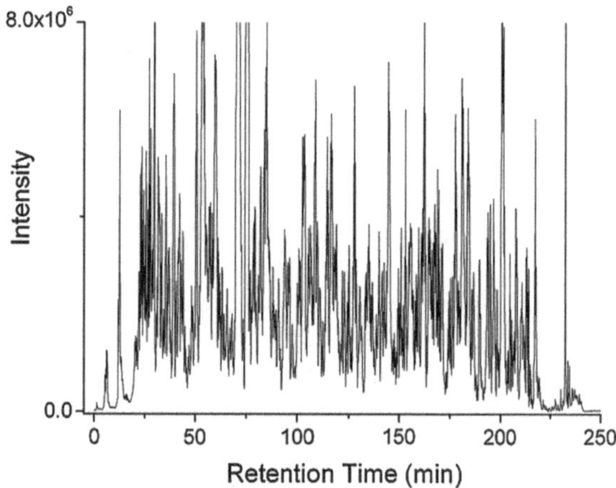

Fig. 3.3 Base peak chromatogram of nano-RPLC-MS/MS analysis of 1 μg yeast protein tryptic digest by 85 cm × 75 μm i.d. C12 RP monolithic column prepared by ternary porogenic solvent

Table 3.2 Column-to-column retention reproducibility of nano-RPLC-MS/MS analysis of tryptic digest of 1 μg yeast proteins on three different columns prepared in different batches at three different days*

Mass	Retention time (min)				RSD (%)
	Column 1	Column 2	Column 3	Average	
1,591.75	38.15	36.86	35.25	36.75	3.2
1,840.92	56.96	55.62	51.99	54.86	3.8
1,373.64	74.48	73.16	66.97	71.54	4.6
1,470.84	82.37	81.55	77.74	80.55	2.5
1,598.79	99.07	98.90	92.14	96.70	3.3
1,815.83	113.60	113.37	107.62	111.53	2.5
2,160.11	123.68	122.42	119.3	121.80	1.5
1,635.85	137.16	136.16	134.58	135.97	0.8
2,744.50	143.44	142.12	141.7	142.42	0.5
2,388.20	159.85	158.62	156.66	158.38	0.8
2,403.23	186.90	182.03	184.01	184.31	1.1
2,387.34	220.35	213.16	213.79	215.77	1.5

* Peptides are randomly selected across the elution, sample loading rate is 15 μL/min, and the separation conditions are the same as Fig. 3.3

the polymerization mixture, but also the constitution of the porogenic solvent is crucial to the homogeneity of the porous monolithic structure, which is also the most important factor to both high-efficient LC separation and proteome analysis.

3.2.1.2 Effect of Functional Monomer

Reversed phase liquid chromatographic separation is the most commonly utilized separation mode in shotgun proteome analysis due to its powerful separation ability. Silica gel particles modified with C18 group are most usually packed into capillary column for RP separation. For methacrylate-based monolith, the hydrophobic groups are introduced by direct copolymerization of corresponding functional monomer. Eeltink et al. optimized the porous structure and polarity of the methacrylate-based monolithic capillary columns, and they demonstrated that C12 monolithic capillary column could finally give the best separation performance [27]. In this section, LMA and stearyl methacrylate (SMA) were also utilized to introduce C12 and C18 groups into the hydrophobic monolithic columns, respectively. The constitute ratio of polymerization mixture LMA:EDMA:1-propanol:1,4-butanediol at 100:100:147:73 and SMA:EDMA:1-propanol:1,4-butanediol at 100:100:133:67 was applied, respectively. Similar permeability was observed for these two types of monolithic columns as shown in Table 3.1. When the 85 cm × 75 μm i.d. C12 monolithic column was applied, 1,787 unique peptides (RSD = 2.3 %, $n = 3$) and 574 proteins (RSD = 0.1 %, $n = 3$) were positively identified from 1 μg tryptic digest of yeast protein in 210-min separation gradient as described above. Then, 1,607 unique peptides (RSD = 2.2 %, $n = 3$) and 567 proteins (RSD = 2.6 %, $n = 3$) were positively identified by the 85 cm × 75 μm i.d. C18 monolithic column at the same condition. Though SMA had higher hydrophobicity than LMA, similar proteome coverage was obtained for these two types of monolithic columns as similar porous property was introduced by the binary porogenic solvent as shown in Fig. 3.1b, d. XICs of the three peptides for the above C12 and C18 monolithic columns were also extracted and shown in Fig. 3.2b, d, and it could be seen that the average peak widths at 0.613 height of the peak was 0.75 and 0.78 min with corresponding peak capacity at 141 and 135, respectively. Therefore, the porous structure might be the most important factor to the separation performance and proteome coverage when similar hydrophobic functional monomers were applied.

3.2.1.3 Effect of Column Length, Separation Gradient, and Online Multidimensional Separation

Three C12 monolithic columns with length of 30, 60, and 85 cm were synthesized within a 75-μm-i.d. capillary using the polymerization mixture with ternary porogenic solvent. When 92-min RP separation gradient was applied, the identified unique peptides and proteins increased from 717 and 343 to 1,435 and 426 during the increase of column length from 30 to 85 cm as shown in Fig. 3.3. Though the identified unique peptides and proteins increased 100.1 and 24.2 % respectively, the operating pressure of the system was also increased from 550 to 1,774 psi. Therefore, the operating pressure of LC instrument limits the increase of column length. The average peak capacities were 142, 135, and 150 for column length at 30, 60, and 85 cm, respectively. The peak capacities were not significantly changed with different column length under the same LC separation gradient. This can be partially explained by the fact that the steepness was changed along with the increasing of the column length as reported by Wang et al. [28].

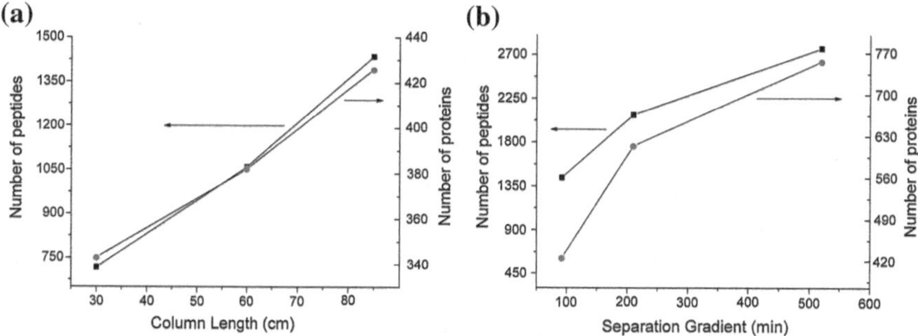

Fig. 3.4 **a** Unique peptides and proteins identified by 30, 60, and 85 cm C12 monolithic columns synthesized by ternary porogenic solvent containing 20 μL water with 92-min separation gradient; **b** unique peptides and proteins identified by 85 cm × 75 μm i.d. C12 monolithic column synthesized by the ternary porogenic solvent containing 20 μL water with 92, 210, and 520-min RP separation gradient (0–30 % buffer B) (Reprinted with permission from Ref. [24]. Copyright 2009 Elsevier)

As complex biological samples are usually analyzed in proteomics, ultra-long RP separation gradient can be utilized to increase the proteome coverage especially for relative long column [29]. Three RP binary separation gradient 92, 210, and 520-min were applied to the 85 cm × 75 μm i.d. C12 monolithic column synthesized by the ternary porogenic solvent. The identified unique peptides and proteins increased from 1,435 and 426 to 2,765 and 756 during the increase of separation gradient from 92 to 520-min as shown in Fig. 3.4. Though the identified unique peptides and proteins increased 92.7 and 77.5 %, the analysis time was also increased 465.2 %. When 210-min separation gradient was applied, 2,081 unique peptides and 614 proteins were positively identified. Though the proteome coverage was a little inferior to 520-min separation gradient (75.3 % unique peptides and 81.2 % proteins), it might be more suitable to this 85-cm-long C12 column considering only 40.4 % separation time was consumed. The XICs of the three peptides were also investigated along with the increase of the separation gradient, and the average peak capacities were 150, 173, and 222 under separation gradient at 90, 210, and 520 min, respectively. These results were consistent with the previous reports that the peak capacity would be increased along with the increasing of separation gradient when the other conditions are the same [29].

Multidimensional separation is one of the most efficient methods to improve the separation efficiency in nanoflow LC-MS/MS analysis. Since the 85 cm × 75 μm i.d. C12 monolithic column synthesized by the ternary porogenic solvent had the most powerful separation capability as described above, it was coupled with a 7 cm × 150 μm i.d. phosphate monolithic trap column described in Chap. 2 to construct an automated proteome analysis system by vented column configuration. And fast flow rate for sample injection (15 μL/min, 4 min) and operation of the whole system was achieved at the pressure only about 1,800 psi. As the phosphate monolith exhibits good SCX mechanism and the C12 monolith has strong hydrophobicity, online multidimensional separation can be easily achieved in this totally monolithic columns system. Twenty microgram tryptic digest of yeast proteins was automatically injected onto the 7 cm × 150 μm i.d. phosphate monolithic trap column. Then, 14 cycles of salt stepwise elution of peptides from phosphate monolithic trap column to C12 monolithic analytical column followed by 210-min gradient separation and MS detection were automatically conducted. Figure 3.5 gives the base peak chromatograms of this online multidimensional LC-MS/MS analysis. It can be seen that the tryptic peptides were symmetrically distributed among the 14 SCX fractions by the salt stepwise elution. After database search as described and controlling false discovery rate (FDR) of peptides identification less than 1 %, 2,051 distinct proteins were positively identified with 8,500 unique peptides (total 78,864 peptides) and the percentage of proteins identified by two or more unique peptides per protein was 60.8 %.

The good separation capability of this online multidimensional separation system can be attributed to two aspects. This first one is the good orthogonality between the phosphate SCX and C12 hydrophobic monoliths, which provides efficient prefractionation in the first dimension of online stepwise salt elution, and the newly discovered unique peptides were symmetrically distributed among 14-steps salt elution (Fig. 3.6). The other one is the good separation performance of the C12 monolithic column. It is difficult to increase the separation capability of organic polymer-based monolithic column due to the inhomogeneity of the overall pore structure. We have demonstrated that the porogenic solvent plays an important role to optimize the porous structure of monolith. When the ternary porogenic solvent including water was used, the overall pore structure on C12 hydrophobic monolith could became more homogeneous, which increased the separation capability greatly. We had also optimized the functional monomer, column length, and separation gradient for the monolithic column, which also contributed to the good separation capability of this online multidimensional separation system. Further, an integrated ESI tip was directly pulled from the end of the separation monolithic capillary column in this study. This type of integrated ESI tip can avoid using the connecting union and eliminate the void volume between separation column and ESI tip, which feasibly increases the LC separation performance. Replicate analysis was also performed for this 14-steps online 2D SCX-RP LC-MS/MS analysis. The base peak chromatograms of cycle 8 were both investigated. Obviously, similar peak intensity and LC elution profiles were obtained in both analyses. The XIC of three randomly selected unique peptides were also checked, and the retention time, peak intensity, and elution profile were all similar (Fig. 3.7). Therefore, this totally monolithic column system is highly reproducible in both LC separation and proteome analysis.

3.2.2 Online 2D LC-MS/MS Analysis by Using Biphasic Monolithic Column

In Multidimensional protein identification technology (MudPIT), the SCX and RP solid phase materials are packed into a single capillary column in sequence and online multidimensional separation can be achieved by stepwise elution of salt

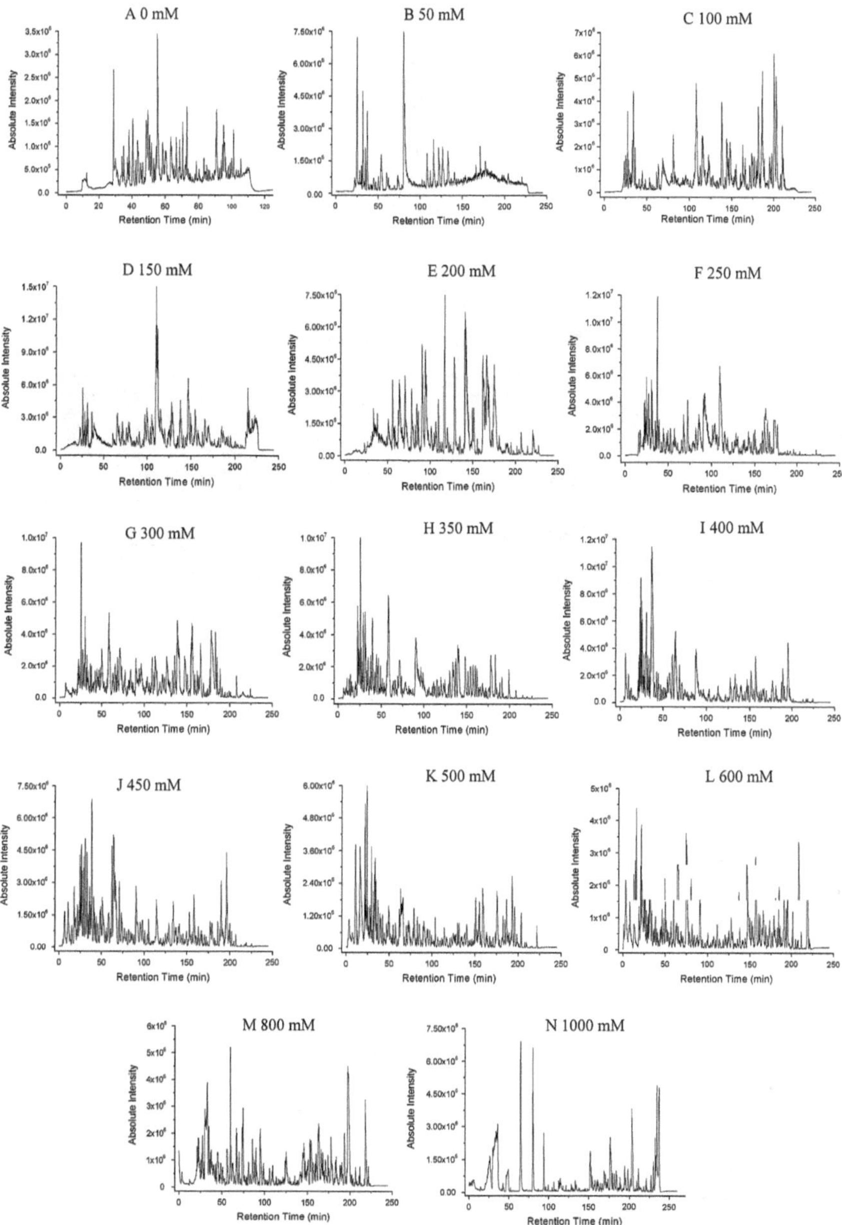

Fig. 3.5 Chromatograms of a 14-cycle online multidimensional analysis of 20 μg tryptic digest of yeast proteins (Reprinted with permission from Ref. [24]. Copyright 2009 Elsevier)

buffer with different concentration from SCX to RP segment [30–32]. However, limited amount of SCX and RP materials can be packed due to the back pressure of the capillary column and the limit of working pressure of LC instrument.

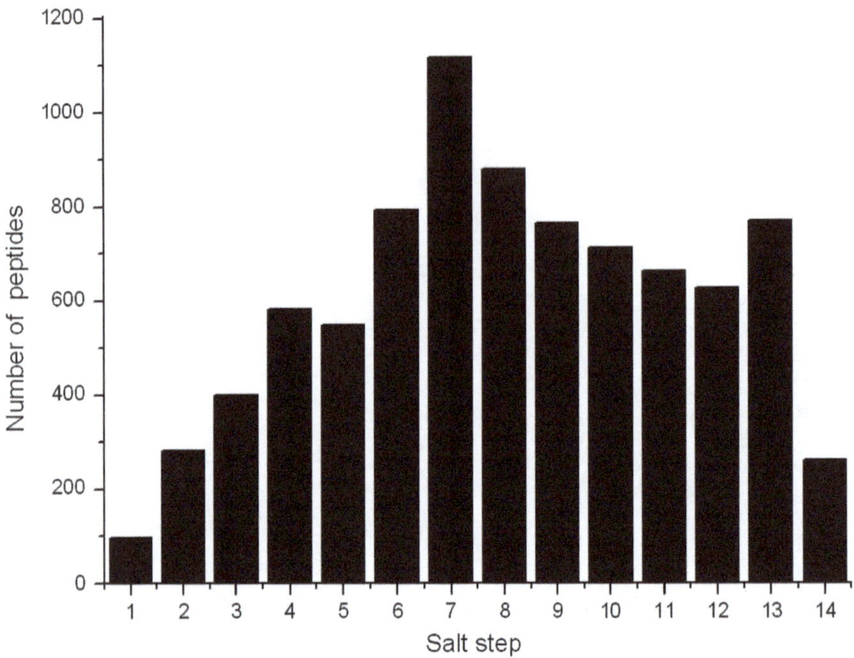

Fig. 3.6 Number of newly unique peptides identified from each of the 14 cycles of the online multidimensional analysis shown in Fig. 3.5

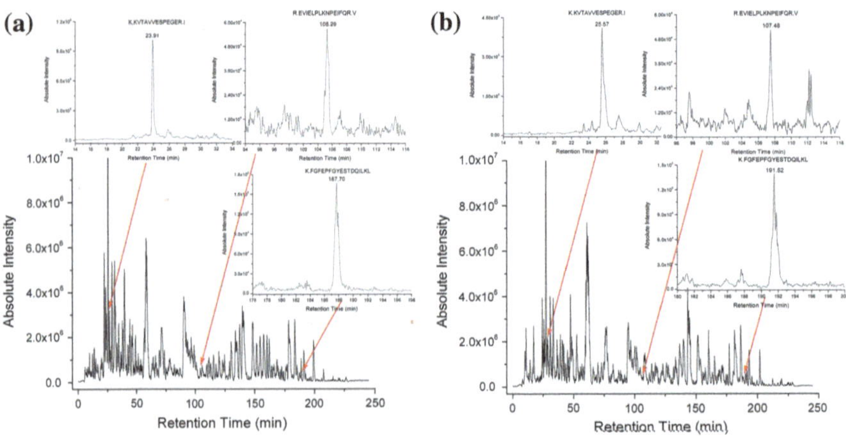

Fig. 3.7 Two base peak chromatograms of the cycle 8 in two online multidimensional LC-MS/MS analysis performed on consecutive days. Extracted ion currents (XICs) of three moderately intense peptides were extracted to estimate the reproducibility of the 2D analysis

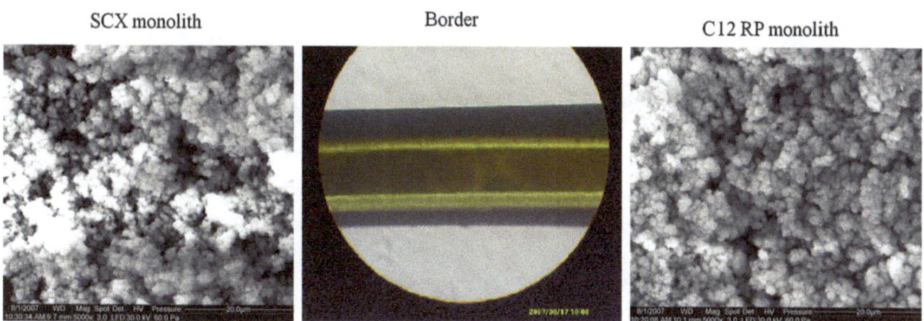

Fig. 3.8 Scanning electron microscopy photographs of the biphasic monolithic column with 100 μm i.d. at magnification of 5,000× for the SCX segment and RP segment. And the border of the two types of monoliths in the biphasic monolithic capillary column (Reprinted with permission from Ref. [33]. Copyright 2008 American Chemical Society)

Monolithic columns exhibit superior permeability than particulate columns, and we have already developed phosphate SCX and C12 hydrophobic monolithic columns with good LC separation performance as described in previous sections. Therefore, we prepared both phosphate SCX and C12 RP monoliths into a single 100-μm-i.d. capillary column.

To eliminate the dead volume between SCX and RP monolithic segments, we first tried to fill a 75-cm-long capillary column with 10-cm-long SCX polymerization mixture and 65-cm-long RP polymerization mixture in sequence by using a syringe followed with simultaneous polymerization of the both sections. Unfortunately, diffusion would occur between the two absolutely different polymerization mixtures sections and the border between them was easy to form a colloid-type polymer, which could not be permeated with solution even at pressure over 6,000 psi. Thus, we have to prepare the SCX and RP monoliths separately in two steps procedures. The scanning electron microscopy photographs of the SCX and RP monolith segments in the biphasic monolithic capillary column are shown in Fig. 3.8a, b, and the border between two types of monoliths is shown in Fig. 3.8c (SCX monolith is on the left). It can be seen that the middle area between the two types of monoliths is a little incompact but the difference is not so obvious. The length of the incompact section is about 1 mm, and the corresponding volume is less than 10 nL. Therefore, the dead volume resulted from the incompact monolith section would be much smaller than that in the connection of two types of monolithic columns even with a ZDV union.

Then, this SCX-RP biphasic monolithic column was applied in MudPIT online 2D LC-MS/MS analysis and the tryptic digest of 10 μg yeast proteins was used as testing sample. A five-cycle online 2D LC-MS/MS analysis was performed in about 12 h including total binary gradient separation time 460 min, which was similar to 1D separation with relatively long separation gradient in some cases (Fig. 3.9) [3, 4]. Finally, 780 distinct proteins were positively identified from 2,953 unique peptides (total 15,512 peptides) with false positive rate at 0.85 % by

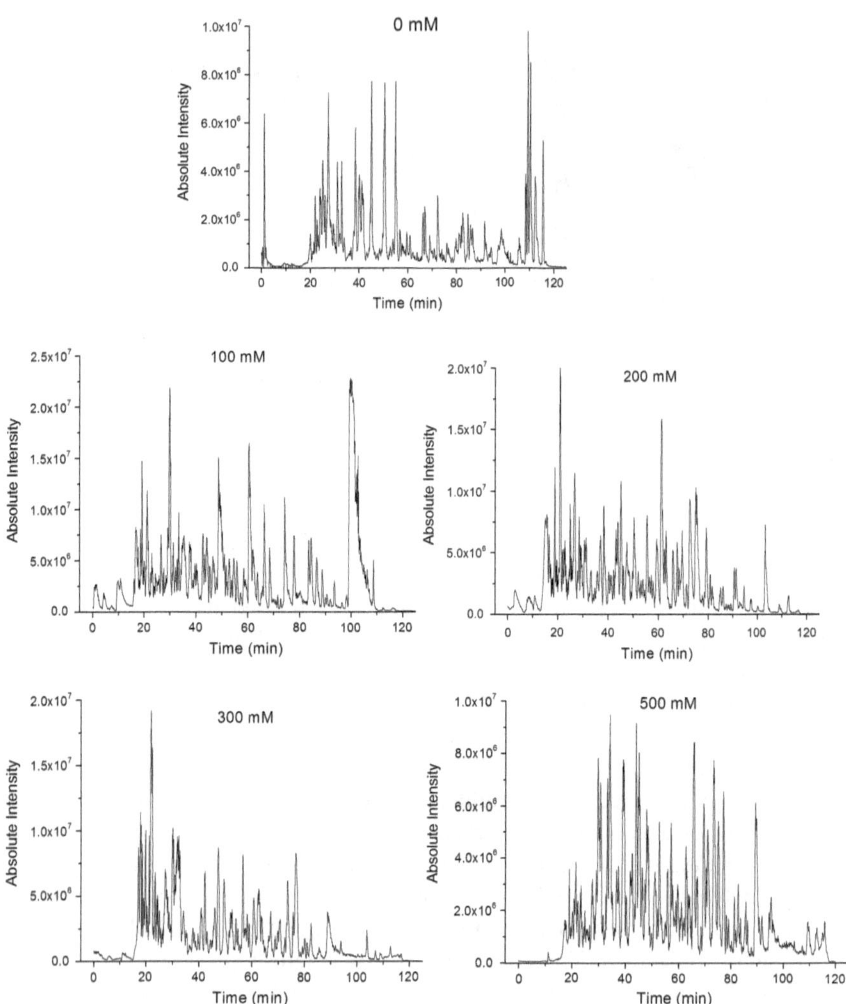

Fig. 3.9 Base peak chromatograms of the five-cycle online multidimensional separation of tryptic digest of 10 μg yeast proteins (Reprinted with permission from Ref. [33]. Copyright 2008 American Chemical Society)

setting ΔC_n at 0.35 (Xcorr: 1.9, 2.2, 3.75 for singly, doubly, triply charged peptides). If we set ΔC_n at 0.19 to control the false positive rate at 4.95 % and only proteins have two or more unique peptides are thought to be positively identified as Motoyama et al. described [34], 579 yeast proteins were obtained. Though this number is less than reported 742 yeast proteins with the same gradient volume 1.4 (gradient length/RP column length, 92/65), the total run time that they applied was 27.5 h and the operating pressure of their system was ~15 kpsi contrasting to 12 h

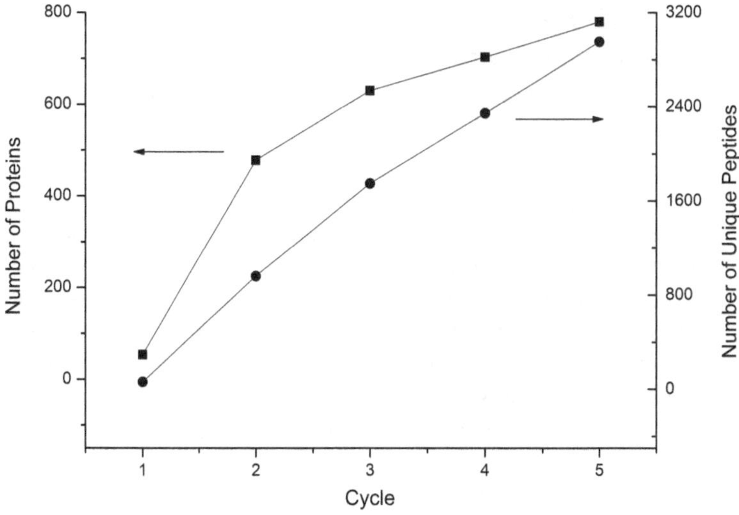

Fig. 3.10 Total number of the newly identified proteins (*left*) and unique peptides (*right*) along with the separation cycles during online multidimensional separation (Reprinted with permission from Ref. [33]. Copyright 2008 American Chemical Society)

and ~900 psi applied in our system [34]. Therefore, good separation performance was obtained by this biphasic monolithic column at much lower LC operating pressure due to the superior permeability of monolithic columns. The numbers of the newly identified proteins and unique peptides along with the salt stepwise elution were shown in Fig. 3.10, which both indicate the good orthogonality of the SCX and RP monoliths in this online 2D LC-MS/MS analysis.

This biphasic monolithic column can be easily prepared and applied in online 2D LC-MS/MS analysis conveniently. The two types of monoliths synthesized in the capillary column have both been demonstrated to be stable enough in the separation buffers for longtime usage as described in previous sections. Comparing to the conventional MudPIT type of biphasic packed column, this biphasic monolithic column has the following advantages. First, the operating pressure is relative low and can be applied in common HPLC systems without any difficulties. In this study, though a 75-cm-long biphasic column was used, the operating pressure was only ~900 psi, which makes the operation of the whole system very easy and robust. Second, as relatively long SCX and RP segments can be applied, the sample loading capacity and separation peak capability can be improved correspondingly [25]. Third, as the biphasic column is easy to prepare, it is possible to prepare this type of biphasic column in capillary with much smaller internal diameter (such as 10 μm), which will increase the sensitivity of nanoflow LC-MS/MS system greatly as the MS is a concentration-dependent detector [35].

3.2.3 Monolithic Electrospray Ionization Emitter and its Application in Label-Free Quantitative Proteome Analysis

As we discussed above, the LC separation performance is crucial to the performance of proteome analyses. Although many efforts have been paid onto the development of new LC separation technologies, such as new solid phase materials and new multidimensional separation systems, eliminating the influences that decrease the LC separation capability is also important. Dead volume within the LC system has significant influence on the LC separation performance due to the additional distortion of gradient separation profile and the remix of separated analytes. There are usually three types of dead volume in LC separation. The first type of dead volume is located in the front of both trap and separation columns, such as the dead volume introduced by the connecting void capillary from the pump to trap column. This type of dead volume will delay the elution profiles in based peak chromatogram and prolong the total analysis time, but has little influence on the LC separation performance. The second type of dead volume is located in the middle of the trap and separation columns. This type of dead volume seriously decrease the separation performance of the LC system and the usage the SCX trap column can effectively eliminate the influence of this type of dead volume as described in Chap. 2. The third type of dead volume is located behind the LC separation column, such as the void ESI emitter and connecting union between the separation column and ESI emitter. The well-separated analytes will inevitably remix within this type of post-column dead volume, which is extremely harmful to the LC separation performance. However, the post-column dead volume has not been well studied until now, and all of the nanoflow electrospray ionization (nano-ESI) emitters in commercial available LC instruments are fabricated by 25-μm-i.d. void capillary. Therefore, in this section, we evaluated the influence of the post-column dead volume introduced by void ESI emitter on the LC separation performance. As the C12 hydrophobic monolith exhibited much higher permeability than packed solid phase material as described above, it was also introduced into the nano-ESI emitter to eliminate the post-column dead volume.

The procedures for preparation of the C12 monolithic emitter are the same as preparation of the C12 monolithic column except for that 7-cm-long C12 monolithic column with integrated ESI tip is cut and used as monolithic emitter as shown in Fig. 3.11. We can prepare an ultra-long C12 monolithic column once (such as 1.5 m) to obtain multi emitters by cutting at every 7-cm length. By comparing to the 25-μm-i.d. commercial available void emitter, the back pressure of the C12 monolithic emitter was only 200–300 psi higher at the same inner diameter due to the high permeability of the C12 monolith. However, the tip of 25-μm-i.d. void emitter is easy to be blocked, which limits its usage time in 1 week at best. For C12 monolithic emitter, tip blocking will not occur due to the introduction of hydrophobic monolith and its usage time can last to about 2 months, which extremely benefit for longtime nanoflow LC-MS/MS analysis such as online multidimensional separation.

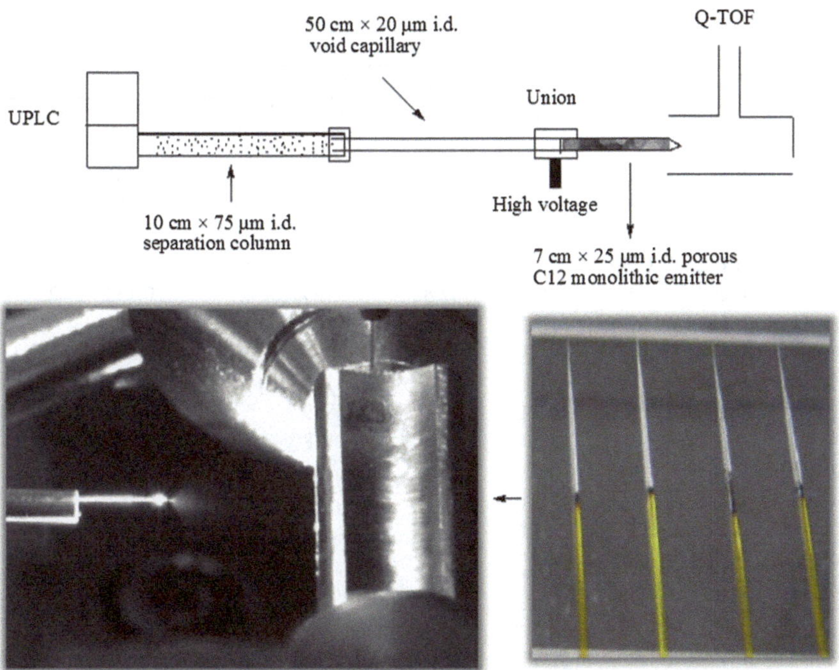

Fig. 3.11 Schematic diagram of the commercial available nano-RPLC-MS/MSE system and picture for homemade porous C12 monolithic ESI emitter (Reprinted with permission from Ref. [36]. Copyright 2008 Wiley)

A 500 fmol tryptic digest of BSA was used as testing sample to investigate the effect of ESI emitter on separation efficiency. The XICs of four peptides were selectively extracted base peak chromatograms and shown in Fig. 3.9. The 75-μm-i.d. void capillary ESI emitter was applied at first, and the obtained XICs of the four peptides, (R)SLGKVGTR(C) (RT 17.78 min, +2), (R)LCVLHEK(T) (RT 23.71 min, +2), (R)KVPQVSTPTLVEVSR(S) (RT 24.03 min, +2), and (R)MPCTEDYLSLILNR(L) (RT 32.55 min, +2) are shown in Fig. 3.9a. It can be seen that the peak profiles of the first three peptides are a little branched; the second and third peptides are not separated completely. When the 75-μm-i.d. C12 monolithic ESI emitter was used, much better elution profiles of the four peptides could be obtained as shown in Fig. 3.9b. As the peak broadening was much less than that by 75-μm-i.d. void emitter, better separation could be achieved for the second and third peptides. Finally, 25-μm-i.d. void and C12 emitters were also tested, and the separation profiles of the four peptides were shown in Fig. 3.9c, d. Obviously, the separation performance was better than 75-μm-i.d. emitters due to the reduction of the void volume. When the 25-μm-i.d. C12 monolithic emitter was used, the smallest peak width was obtained and baseline separation of the second and third peptides could be achieved.

Table 3.3 Comparison of $W_{0.613}$ for the extracted peptide peaks by using different types of emitters

m/z	75 μm void emi.		75 μm C12 emi.		25 μm void emi.		25 μm C12 emi.	
	$W_{0.613}$	RSD (%)	$W_{0.613}$	RSD (%)	$W_{0.613}$	RSD (%)	$W_{0.613}$	RSD (%)
732.8	0.135	21.9	0.089	7.1	0.065	5.8	0.060	23.7
409.7	0.157	14.0	0.140	1.7	0.127	4.5	0.101	5.8
740.3	0.199	23.9	0.207	3.9	0.187	0.9	0.162	3.1
752.3	0.208	4.3	0.178	4.5	0.153	10.8	0.127	3.1
875.8	0.192	0.7	0.133	3.3	0.104	7.4	0.101	0.3
598.8	0.208	7.1	0.161	2.8	0.159	4.7	0.133	2.5
488.2	0.157	11.3	0.112	13.6	0.102	4.0	0.095	4.2
653.9	0.164	8.2	0.111	4.7	0.094	4.7	0.087	4.1
450.2	0.191	8.9	0.155	4.4	0.138	11.2	0.121	3.1
821.0	0.180	7.8	0.111	2.8	0.089	5.4	0.086	5.0
941.5	0.092	21.7	0.126	3.9	0.118	9.0	0.109	1.5
710.8	0.161	8.4	0.120	4.6	0.106	5.4	0.086	5.7
740.9	0.183	5.1	0.148	4.2	0.134	3.5	0.113	2.9
700.8	0.191	4.9	0.171	0.9	0.155	3.3	0.140	3.9
863.4	0.194	2.7	0.156	0.9	0.142	7.3	0.120	2.4
Average	0.174	10.1	0.141	4.2	0.125	5.9	0.109	4.8

Reprinted with permission from Ref. [36]. Copyright 2008 Wiley
Notes The $W_{0.613}$ was the average value of three consecutive analyses

The capability of LC separation is usually measured by peak capacity, which is determined by the equation: $nc = L/(2W_{0.613})$, where L is the total time over which peptides elute and $W_{0.613}$ is the peak width at 0.613 height of the peak [5]. Therefore, the $W_{0.613}$ values of 15 peptides randomly selected across separation window were calculated for the four different ESI emitters (Table 3.3). Obviously, C12 emitter systems have narrower peak width and so better separation performance were obtained at the same inner diameter. Besides, the RSDs for the C12 emitter systems are also smaller than that of void emitter systems, which demonstrate that more reproducible results can be obtained. Comparing to the void emitters, the average $W_{0.613}$ values of the 15 peptides have decreased 19.0 and 12.8 % by the 75-μm-i.d. and 25-μm-i.d. C12 monolithic emitters, which also means the peak capacity has increased 19.0 and 12.8 %, respectively. Thus, the ESI emitter has an important effect on improving the separation performance in LC-MS/MS system due to the elimination of post-column dead volume. The strong hydrophobicity of the porous C12 monolith may also play important role in refining the elution profiles of the analytes (Fig. 3.12).

As C12 monolithic emitter improves the separation performance of LC-MS/MS system, it may also improve the accuracy of label-free quantification as the accuracy strongly relies on the quality and reproducibility of LC separation. Two MPDS mixtures (Waters) were spiked into equal amount of tryptic digest of mouse livers extract to prepare the testing sample of label-free quantification. The 25-μm-i.d. commercial available void emitter (Thermo) and homemade C12

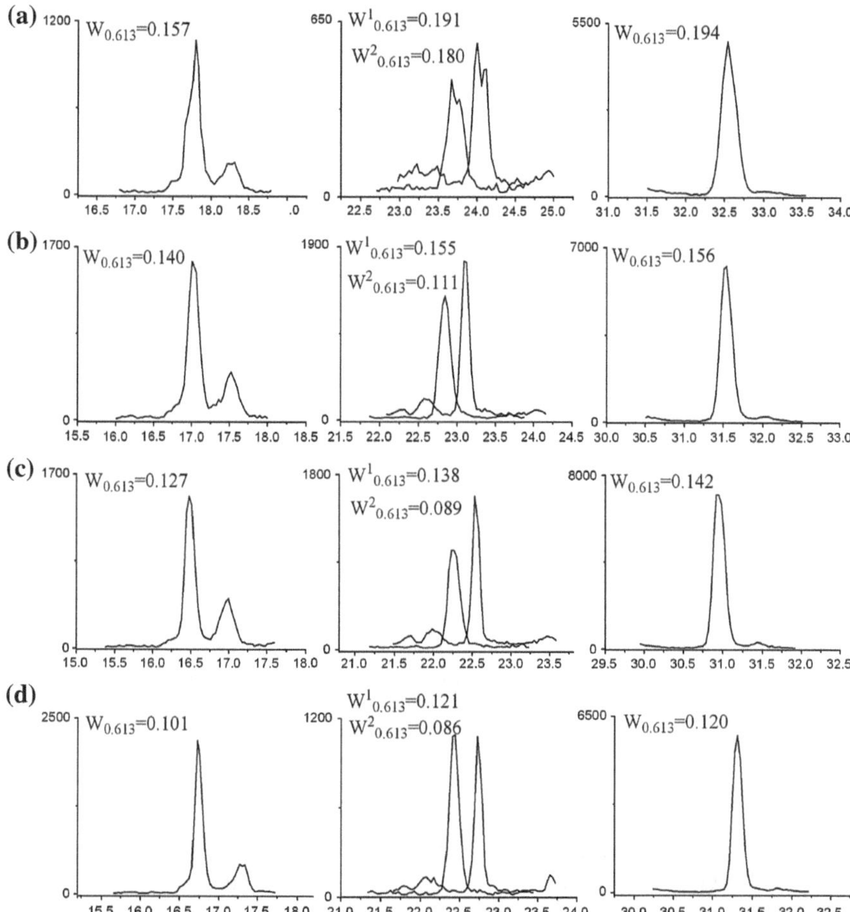

Fig. 3.12 XICs for four peptides [(R)SLGKVGTR(C) (RT 17.78 min, +2), (R)LCVLHEK(T) (RT 23.71 min, +2), (R)KVPQVSTPTLVEV- SR(S) (RT 24.03 min, +2), and (R)MPCTEDYLSLILNR(L) (RT 32.55 min, +2)] by using **a** 75-μm-i.d. void capillary emitter, **b** 75-μm-i.d. C12 monolithic emitter, **c** 25-μm-i.d. void capillary emitter, **d** 25-μm-i.d. C12 monolithic emitter ($W_{0.613}$, peak width at 0.613 height of the peak, which was an average value of three replicate injections) (Reprinted with permission from Ref. [36]. Copyright 2008 Wiley)

monolithic emitter were compared in the label-free quantification system (Waters) for relative quantification of the four standard proteins from the MPDS mixtures in the complex matrix of mouse livers extract.

At first, relative quantification of mixtures 1 and 2 was conducted for void emitter by the Expression Analysis tool embedded in the ProteinLynx Global Server (PLGS) v2.3 [37–40]. Though not quantitative filtering criteria were used, 10,717 matched Exact Mass and Retention Time (EMRT) clusters and 312 proteins could give quantification information as shown in Fig. 3.13a, c. Figure 3.11a

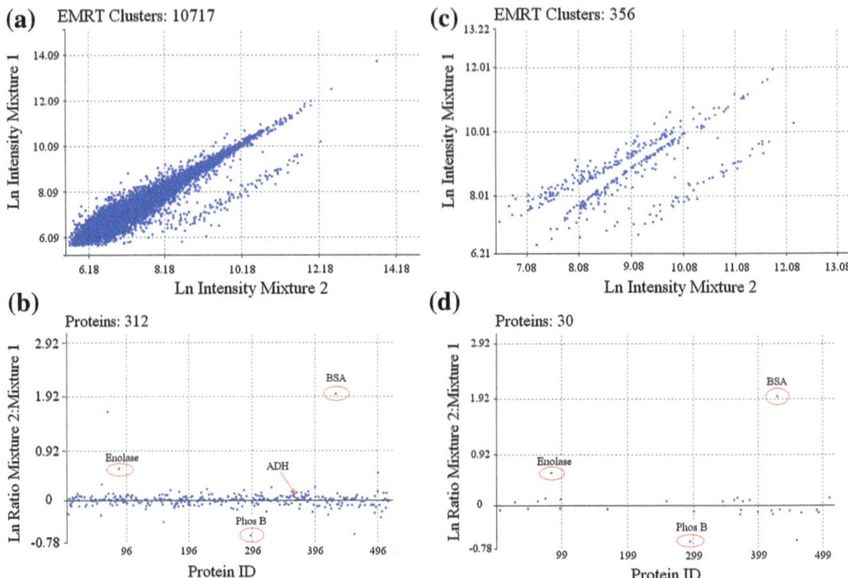

Fig. 3.13 Binary comparison of log intensity of the matched EMRT clusters **a** without filtering criteria, **b** with filtering criteria, and relative quantification of identified proteins in mixture 1 and mixture 2, **c** without filtering criteria, **d** with filtering criteria in the separation system using 25-μm-i.d. void capillary emitter. The filtering criteria were set as: identification in at least two out of three injections and $P \geq 0.95$ for upregulations and ≤ 0.05 for downregulations for both EMRT clusters and proteins (Reprinted with permission from Ref. [36]. Copyright 2008 Wiley)

shows the scatter plot of the resulting binary comparison for the log intensity of matched EMRT clusters in mixtures 1 and 2. It can be seen that most of the EMRT plots distribute along with the diagonal line, which implies that intensity of most EMRT clusters nearly have not changed due to the amount of protein matrix in the two mixtures is the same. However, some clusters were deviated from others and parallel with the diagonal line. These EMRT clusters mainly belong to BSA, the concentration of which is changed eightfolds between mixtures 1 and 2. Figure 3.13c shows the log intensity ratio of the two mixtures. Similarly, the log ratio for most proteins is about zero, which indicates that the concentration of most proteins is the same. When the quantitative filtering criteria were applied (identifications achieved in at least two out of three injections and $P \geq 0.95$ for upregulations and ≤ 0.05 for downregulations), only 356 matched EMRT clusters and 30 identified proteins with reliable quantitative information in relative quantification can be obtained as shown in Fig. 3.13b, d. Only three out of the four standard proteins had given the quantitative information with high P value and the ratio of BSA, enolase, and phos B was 7.17, 1.80, and 0.52 (theoretically is 8, 2, and 0.5) between mixtures 2 and 1 (shown in Fig. 3.13d). The ratio of ADH was not given because the P value was just 0.85, which was lower than 0.95. The corresponding peptides with high P value (pass through the same quantitative filtering

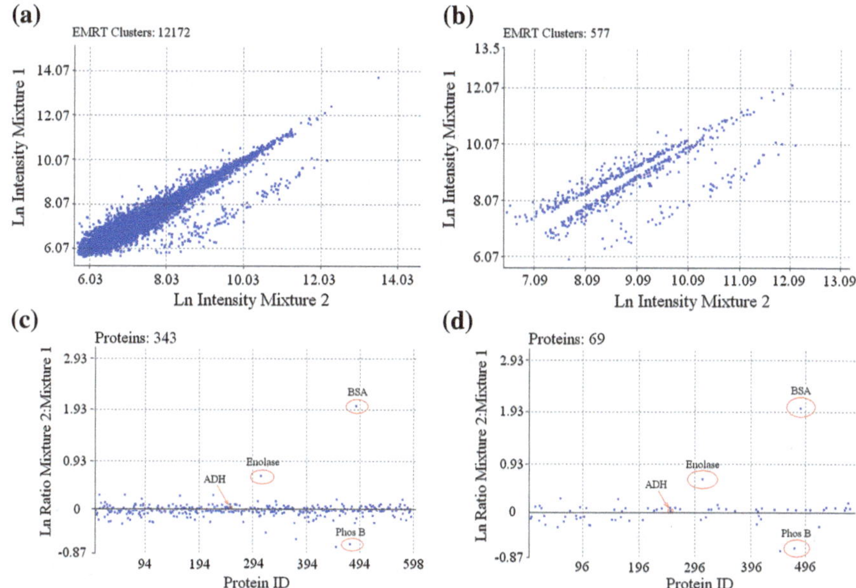

Fig. 3.14 Binary comparison of log intensity of the matched EMRT clusters **a** without filtering criteria, **b** with filtering criteria, and relative quantification of identified proteins in mixture 1 and mixture 2, **c** without filtering criteria, **d** with filtering criteria in the system using 25-μm-i.d. C12 monolithic emitter. The filtering criteria were the same as shown in Fig. 3.13 (Reprinted with permission from Ref. [36]. Copyright 2008 Wiley)

criteria) were also examined and only five peptides belonged to ADH. Excluding the three standard proteins, the average intensity ratio of the 27 mouse liver proteins was 0.98 (theoretically is 1) with RSD at 13.6 %, which validated the reliability of this label-free quantitative strategy.

Then, relative quantification of mixtures 1 and 2 was also conducted for C12 monolithic emitter at the same way. When no quantitative filtering criteria were used, 12,172 matched EMRT clusters and 343 identified proteins can give quantitative information as shown in Fig. 3.14a, c. The total numbers of EMRT clusters and proteins are increased 13.6 and 9.9 % comparing to void emitter system. Furthermore, 577 matched EMRT clusters and 69 identified proteins with high *P* value were obtained as shown in Fig. 3.14b, d when the same quantitative filtering criteria were applied. By comparing to the results obtained from void emitter system, the numbers of matched EMRT clusters and identified proteins were increased 62.1 and 130.0 %, respectively. Additionally, the quantitative information of all four standard proteins with high *P* value can be obtained with the ratio of ADH, BSA, enolase, and phos B at 1.04, 7.46, 1.92, and 0.51 (theoretically is 1, 8, 2, and 0.5) between mixtures 2 and 1, respectively. The *P* value for quantification of ADH was 0.96 and the number of corresponding peptides with high *P* value for ADH was 9. Excluding the four standard proteins, the average intensity ratio of the 64 mouse liver proteins and trypsin was 1.00 (theoretically is 1) with RSD at 14.1 %.

In conclusion, the disruption of the elution profiles in the void emitter would decrease the reproducibility and confidence in the identification of proteins, which also led to the poor results in label-free quantification of proteins. When the C12 monolithic emitter was used, as we had described, the average peak capacity had increased 12.8 % for 25-μm-i.d. emitter, the number of proteins with reliable quantitative information was increased 130.0 %. Therefore, this new type of C12 monolithic ESI emitter might be more suitable to improve the performance of both qualitative and quantitative proteome analyses.

3.2.4 Packed Capillary Column with Integrated Monolithic Frit

Particulate capillary columns packed with uniform silica gel are most widely used in current shotgun proteomics. However, the packing process of particulate capillary column is difficult to control, and the tip of integrated emitter is easy to be blocked by the packing materials. When smaller silica gel materials are utilized to further improve the LC separation performance, the packing process become even harder. Therefore, we prepared a 1-cm-long C12 monolithic frit onto one end of a void capillary column, followed with solid phase packing to prepare the integrated monolithic frit particulate capillary column (IMFPC). Finally, an integrated monolithic nano-ESI emitter was fabricated from the monolith frit (Fig. 3.15).

In order to evaluate the performance of IMFPC column packed with 3 μm ODS-AQ (C18) particles, tryptic digest of 1 μg mouse liver extract was used as testing sample for nanoflow LC-MS/MS analysis (Fig. 3.16a). Then, the XICs of the six peptides, R.VSQEHPVVLTK.F, K.VITAFNEGLK.N, K.INEAFDLLR.S, R.GFLDVVAALR.W, K.AFAISGPFNVQFLVK.G, and K.GEFQILLDALDKIK.T were extracted from the base peak chromatogram and the peak width at 0.613 height of the six peaks ($W_{0.613}$) is varied from 0.13 to 0.27 min (average value of three consecutive analyses, Table 3.4). The average value of the $W_{0.613}$ for all six peaks by three consecutive analyses was determined at 0.22 min. The average peak capacity of the IMFPC column packed with 3 μm ODS-AQ particles calculated by average value of the peak width is 210 under 92-min gradient elution (0–35 % ACN) [5].

Similarly, a fritless capillary column packed with 5 μm ODS-AQ particles was also applied to the separation of tryptic digest of 1 μg mouse liver extract (Fig. 3.16b). It can be seen that the peak profile of tryptic digest of mouse liver proteins on IMFPC column (shown in Fig. 3.16a) was narrower and sharper than that on the fritless capillary column due to smaller packing materials can be packed without blocking the ESI emitter tip. XICs of the same six peptides separated on the fritless capillary column were extracted from the base peak chromatogram and $W_{0.613}$ varied from 0.22 to 0.35 min (Table 3.4), and the average value of all the six peptides is 0.30 min. Then the average peak capacity calculated for fritless capillary column is 154.

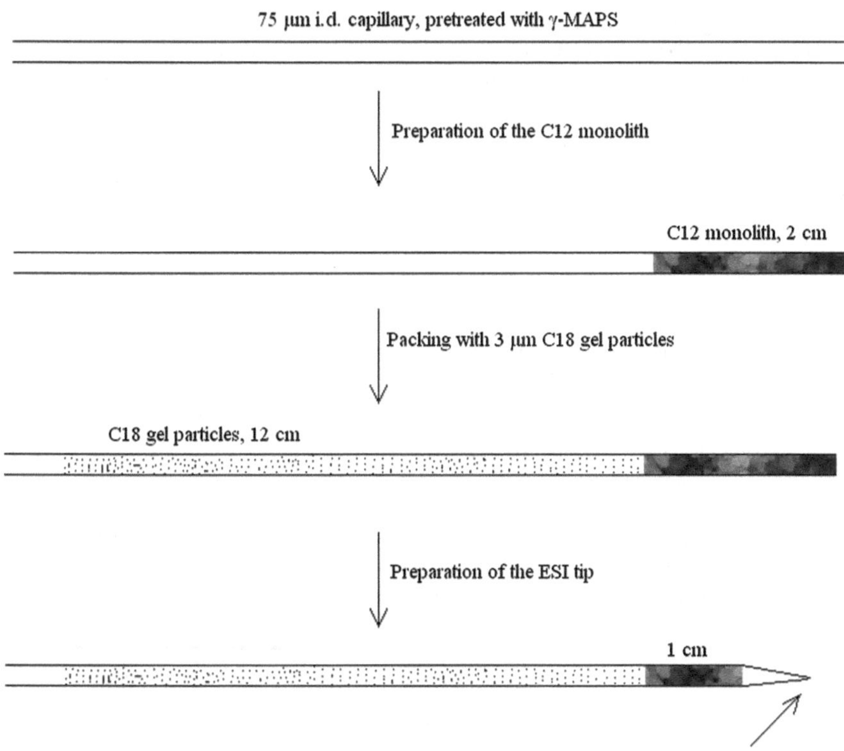

Fig. 3.15 Schematic diagram of the preparation of the IMFPC column (Reprinted with permission from Ref. [41]. Copyright 2009 Elsevier)

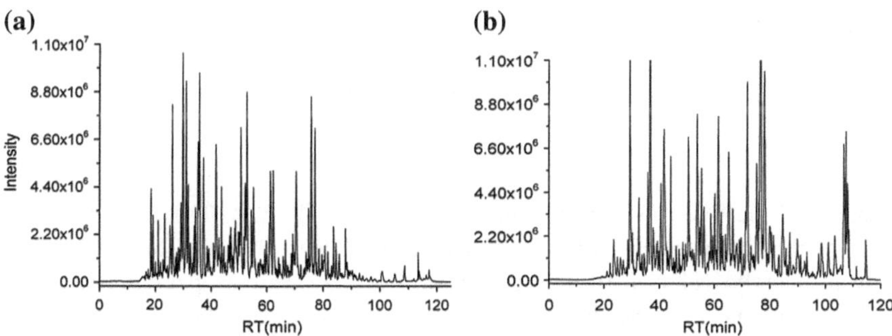

Fig. 3.16 Base peak chromatograms of μLC-MS/MS analysis of 1 μg mouse liver extract by **a** IMFPC column packed with 3 μm ODS-AQ particles and **b** fritless column packed with 5 μm ODS-AQ particles (Reprinted with permission from Ref. [41]. Copyright 2009 Elsevier)

The column-to-column reproducibility was also investigated for the two types of columns. The average $W_{0.613}$ of the six peptides on three IMFPC columns varied from 0.14 to 0.29 with an average value at 0.24. Therefore, the average peak

Table 3.4 Run-to-run reproducibility of $W_{0.613}$ for six peptides R.VSQEHPVVLTK.F, K.VITAFNEGLK.N, K.INEAFDLLR.S, R.GFLDVVAALR.W, K.AFAISGPFNVQFLVK.G, and K.GEFQILLDALDKIK.T by using classical fritless capillary column and IMFPC column

Peptides		1	2	3	4	5	6
IMFPC column	Run 1	0.13	0.22	0.18	0.26	0.27	0.23
	Run 2	0.13	0.19	0.24	0.25	0.23	0.29
	Run 3	0.13	0.18	0.21	0.24	0.26	0.29
	Average	0.13	0.20	0.21	0.25	0.25	0.27
	RSD (%)	0.00	10.58	14.29	4.00	8.22	12.83
Fritless column	Run 1	0.28	0.24	0.33	0.36	0.33	0.23
	Run 2	0.24	0.31	0.34	0.33	0.35	0.28
	Run 3	0.15	0.28	0.32	0.35	0.36	0.32
	Average	0.22	0.27	0.33	0.35	0.34	0.28
	RSD (%)	29.81	12.69	3.03	4.41	4.41	16.30

$W_{0.613}$, peak width at 0.613 height of the peak

Table 3.5 Column-to-column reproducibility of $W_{0.613}$ for six peptides R.VSQEHPVVLTK.F, K.VITAFNEGLK.N, K.INEAFDLLR.S, R.GFLDVVAA- LR.W, K.AFAISGPFNVQFLVK.G, and K.GEFQILLDALDKIK.T by using classical fritless capillary column and IMFPC column

Peptides		1	2	3	4	5	6
IMFPC column	Column 1	0.13	0.22	0.18	0.26	0.27	0.23
	Column 2	0.14	0.22	0.28	0.20	0.28	0.24
	Column 3	0.15	0.24	0.20	0.42	0.27	0.37
	Average	0.14	0.23	0.22	0.29	0.27	0.28
	RSD (%)	7.14	5.09	24.05	38.77	2.11	27.89
Fritless column	Column 1	0.28	0.24	0.33	0.36	0.33	0.23
	Column 2	0.20	0.21	0.19	0.16	0.37	0.17
	Column 3	0.27	0.12	0.32	0.44	0.28	0.30
	Average	0.25	0.19	0.28	0.32	0.33	0.23
	RSD (%)	17.44	32.87	27.89	45.07	13.80	27.88

capacity of the three IMFPC columns was 193 (Table 3.5). The average $W_{0.613}$ value of the three fritless columns varied from 0.19 to 0.33 for the six peptides with an average value at 0.27 (Table 3.5). Thus, an average peak capacity 173 could be obtained for the three fritless columns. Moreover, the average RSD of $W_{0.613}$ values for the six peptides in three columns' analysis was 17.51 % for IMFPC columns and 27.49 % for fritless columns, which demonstrated that better column-to-column reproducibility could be obtained for IMFPC columns.

When IMFPC column packed with 3 μm ODS-AQ particles was applied, totally 640 proteins and 2,287 unique peptides could be obtained in three consecutive analyses of 1 μg tryptic digest of mouse liver protein, among which the percentages of proteins and peptides identified at least two or three times were 67.5,

48.6 % and 63.4, 39.3 %, respectively. When fritless capillary column packed with 5 μm ODS-AQ particles was used, totally 537 proteins and 1,972 unique peptides were obtained in three consecutive analyses. The percentages of proteins and peptides identified at least two or three times were 67.0, 47.7 % and 60.6, 35.0 %, respectively. The column-to-column reproducibility in shotgun proteome analysis was also investigated for the two types of columns. When three IMFPC columns were applied, totally 672 distinct proteins and 2,382 unique peptides were identified by the three IMFPC columns, among which the percentages of proteins and peptides identified by at least two and three columns were 64.1, 44.2 % and 61.0, 36.7 %, respectively. When three fritless capillary columns packed with 5 μm ODS-AQ particles were used, totally 585 distinct proteins and 2,110 unique peptides were obtained by the three fritless capillary columns, among which the percentages of proteins and peptides identified by at least two and three columns were 62.0, 41.5 % and 56.2, 30.2 %, respectively.

The main advantage of IMFPC column is the easiness for packing of smaller solid phase materials. As a monolithic frit is integrated at one end of the capillary, the packing process becomes much easier and faster comparing to fritless capillary column due the high permeability of the monolithic frit. Furthermore, smaller packing particles (less than 5 μm) could be easily packed and would not be clogged during the packing process, which increases the separation capability evidently as described above.

3.2.5 Conclusion

In conclusion, although methacrylate-based monolithic column has already been widely utilized in CEC for high-efficient separation of different types of analytes, they are rarely applied in large-scale proteome analyses. In this chapter, we investigate the different applications of C12 LMA methacrylate-based monolithic column in large-scale proteome analysis. To optimize the separation performance of C12 monolithic column for peptides mixture, both the constitution of polymerization mixture and LC separation conditions were investigated. Further, we established an online 2D LC-MS/MS system by coupling phosphate SCX and C12 RP monolithic columns, and both high separation capability and high proteome coverage were obtained due to the good orthogonality of the two monolithic columns. We can also prepare both SCX and RP monoliths into a single capillary column with relative long column length to obtain a biphasic monolithic column for high-performance MudPIT analysis. Finally, the C12 monolithic column can be also used to fabricate nano-ESI emitter and frit of particulate column to further improve the separation capability of the whole LC system. Therefore, the organic polymer-based monolithic column has various applications in comprehensive proteome analyses to improve the performance in both proteome identification and quantification.

3.3 Experimental Section

3.3.1 *Materials*

Fused silica capillaries with 75 and 150 μm i.d. were purchased from Polymicro Technologies (Phoenix, AZ, USA), and 50 μm i.d. from Yongnian Optical Fiber Factory (Hebei, China). All the water used in experiments was purified using a Mill-Q system from Millipore Factory (Bedford, MA, USA). LMA, SMA, EDMA, and γ-methacryloxypropyltrimethoxysilane (γ-MAPS) were all obtained from Sigma (St. Louis, MO, USA). Azobisisobutyronitrile (AIBN) was obtained from Shanghai Fourth Reagent Plant (Shanghai, China). Trypsin was obtained from Promega (Madison, WI, USA). Dithiothreitol (DTT), iodoacetamide, and protease inhibitors cocktail were all purchased from Sino-American Biotechnology Corporation (Beijing, China).

3.3.2 *Sample Preparation*

The yeast protein extract was prepared in a denaturing buffer containing 50 mM Tris/HCl (pH 8.1) and 8 M urea. The protein concentration was determined by BCA assay. The protein sample was reduced by DTT at 37 °C for 2 h and alkylated by iodoacetamide in dark at room temperature for 40 min. Then the solution was diluted to 1 M urea with 50 mM Tris/HCl (pH 8.1). Finally, trypsin was added with weight ratio of trypsin to protein at 1/25 and incubated at 37 °C overnight. Then, the tryptic digest was purified with a homemade C18 solid phase cartridge and exchanged into 0.1 % formic acid water solution. Finally, the samples were stored at −20 °C before usage. The preparation of mouse liver protein sample was similar to the yeast protein sample.

3.3.3 *Preparation of Monolithic Capillary Columns*

Prior to the polymerization, the capillary was pretreated with γ-MAPS. The methacrylate-based monolithic capillary column was prepared as follows: Briefly, appropriate amount of lauryl or SMA, EDMA, 1-propanol, 1,4-butanediol, and water was mixed with 2 mg AIBN, sonicated for 20 min to obtain a homogeneous solution. After the pretreated capillary was completely filled with the mixture, it was sealed at both ends with rubber stoppers. The sealed capillary was submerged into a water bath and allowed to react for 12 h at 60 °C. The resultant monolithic capillary column was washed with methanol using a HPLC pump to remove unreacted monomers and porogens. Finally, the C12 monolithic capillary column was flushed with water and an integrated ESI tip was directly pulled from the end of column.

For the preparation of phosphate monolithic capillary column, 80 μL ethylene glycol methacrylate phosphate, 60 mg methylene bis-acrylamide, 270 μL dimethylsulfoxide, 200 μL dodecanol, 50 μL N,N′-dimethylformamide, and 2 mg AIBN were mixed and the following procedures were the same as preparation of the C12 monolithic capillary column.

3.3.4 Preparation of the C12 Monolithic ESI Emitter and IMFPC Column

The C12 monolithic capillary column was flushed with water and an integrated tip was directly pulled from the end of the column by a butane torch [5]. The inside diameter of the tip was controlled at about 5 μm for 75-μm-i.d. monolithic column and about 10 μm for 25-μm-i.d. monolithic column under the microscope for back pressure consideration. Finally, 7-cm-long monolithic capillary column with the integrated tip was cut off and it was used as porous monolithic ESI emitter.

For comparison, the void empty capillary ESI emitter of 75 μm i.d. was prepared by directly pulling a 5 μm tip from the capillary and cutting at a length of 7 cm. The void capillary ESI emitter of 25 μm i.d. (7-cm length and 10 μm tip) was purchased from ThermoFinnigan (San Jose, CA).

After the pretreated capillary was filled with ~2 cm of the C12 polymerization mixture by siphon, it was sealed at both ends with rubber stoppers to prepare the IMFPC column. The sealed capillary was submerged into a water bath and allowed to react for 12 h at 60 °C. The resultant capillary column with ~2-cm C12 monolith was washed with methanol using an LC pump to remove unreacted monomers and porogens followed with drying by nitrogen gas. Then, the ODS-AQ materials suspended in methanol were packed into capillary column from the open end and retained in capillary with the 2-cm C12 monolith. When 12-cm ODS-AQ materials were packed, the capillary column was connected to a LC pump and flushed with water at 3,000 psi for 30 min. Finally, an integrated electrospray emitter tip was directly pulled from the end of 2-cm C12 monolith by a flame and only ~1-cm C12 monolith was remained.

3.3.5 Sample Injection and Chromatographic Separation

The LC system (ThermoFinnigan, San Jose, CA) consisted of a degasser, a quaternary Surveyor MS pump. The three buffer solutions used for the quaternary pump were 0.1 % formic acid water solution (buffer A), ACN with 0.1 % formic acid (buffer B), 1,000 mM NH_4Ac at pH 2.7 (buffer C). The flow rate after splitting was about 200 nL/min. A 92 or 210-min binary gradient with buffer A and buffer B was adopted for RP separation. The 92-min binary separation gradient was from 0 to 8 % buffer B for 2 min, from 8 to 30 % for 90 min, and from 30 to 80 % for 5 min.

After flushing 80 % buffer B for 10 min, the separation system was equilibrated by buffer A. The 210-min binary separation gradient was 0 to 8 % buffer B for 5 min, from 8 to 30 % for 205 min, and the rest was the same as 92-min gradient.

The configuration for the automated sample injection RPLC-MS/MS system was the same as Chap. 2 with exception of the 12 cm × 75 μm i.d. packed C18 column was replaced by 85 cm × 75 μm i.d. C12 monolithic column. During sample injection, the switching valve was switched to close the splitting flow and the flow through from the trap column was switched to waste. Sample was automatically injected at a high flow rate to ensure three sample volumes of buffer A was passed through the trap column for removing contaminants. In online multidimensional separation, 20 μg tryptic digest of yeast protein was automatically loaded onto the phosphate monolithic trap column. A series stepwise elution with salt concentrations of 50, 100, 150, 200, 250, 300, 350, 400, 450, 500, 600, 800, and 1,000 mM NH_4Ac was used to gradually elute peptides from phosphate monolithic column onto the 85-cm-long C12 monolithic column. Each salt step lasts 5 min except last one for 15 min. After whole system was re-equilibrated for 10 min with buffer A, 210-min binary separation gradient as described above was applied to separate peptides prior to MS detection in each cycle.

3.3.6 *Mass Spectrometry Detection*

The LTQ linear ion trap mass spectrometer equipped with a nanospray source, a 6-port/two-position valve (Thermo, San Jose, CA). The temperature of the ion transfer capillary was set at 200 °C. The spray voltage was set at 1.82 kV and the normalized collision energy was set at 35.0 %. One microscan was set for each MS and MS/MS scan. All MS and MS/MS spectra were acquired in the data dependent mode. The mass spectrometer was set that one full MS scan was followed by six MS/MS scans on the six most intense ions. The dynamic exclusion function was set as follows: repeat count 2, repeat duration 30 s, and exclusion duration 90 s. System control and data collection were done by Xcalibur software version 1.4 (Thermo).

The nano-UPLC was coupled with Q-TOF Premier MS (Waters) through different types of ESI emitters. For all measurements, the MS was operated in V-mode with typical resolving power of at least 8,000. The MS detector was calibrated from m/z 50 to 1,950 by the MS/MS fragment ions of GFP. All analyses were performed using positive-mode ESI and the spectrum integration time was 0.9 s with an interscan delay time of 0.1 s. Accurate mass LC/MS and LC/MSE data were collected using 8 eV for MS and 28–35 eV for MSE acquisition, such that one cycle of MS and MSE data was acquired every 2.0 s. The radio frequency offset applied to the quadrupole mass analyzer was adjusted such that the LC/MS data were effectively acquired from m/z 300 to 2,000. Mass accuracy was maintained throughout the analysis time due to the usage of a LockSpray. GFP (150 fmol/μL) was used as reference, which was delivered through the reference sprayer and scanned every 30 s.

3.3.7 Data Analysis

The acquired MS/MS spectra were searched against the database using the Turbo SEQUEST in the BioWorks 3.2 software suite (Thermo, San Jose, CA). In order to evaluate the FDR, corresponding reversed sequences were appended to the yeast protein database, which have 5,883 entries. Cysteine residues were searched as static modification of 57.0215 Da, and methionine residues as variable modification of 15.9949 Da. Peptides were searched using fully tryptic cleavage constraints and up to two internal cleavages sites were allowed for tryptic digestion. The mass tolerances were 2 Da for parent masses and 1 Da for fragment masses. The peptides were considered as positive identification if the Xcorr were higher than 1.9 for singly charged peptide, 2.2 for doubly charged peptide, and 3.75 for triply charged peptides. ΔC_n cutoff values were set to control the FDR of peptides identification <1 %, determined by the calculation based on the reversed database.

For label-free proteome quantification, continuum LC-MS/MSE data were processed and searched using ProteinLynx Global Server (PLGS) v2.3 (Water). Protein identifications were obtained with the embedded ion accounting algorithm of the software and by searching the Mouse International Protein Index protein database (v3.17) to which data from ADH, phos B, BSA, and enolase were appended. The ion detection, clustering, and normalization were processed using PLGS. The mass tolerances were 10 ppm for precursor ions and 0.1 Da for fragment ions. Positive identification of the proteins was based on the detection of more than two fragment ions per peptide; more than five fragment ions and two peptides measured per protein. The false positive rate of peptide identification in the ion accounting identification algorithm is set at 4 % with a randomized database. The peptides clustering and protein quantification were conducted by the tool of Expression Analysis embedded in the PLGS v2.3. Prior to conducting quantitative analysis, the observed intensity measurements were normalized using the recommended auto normalization tool. And the results of quantitative comparison were filtered by the identifications achieved in at least two out of three injections as well as quantitative probability (P) value ≥ 0.95 for upregulations and ≤ 0.05 for downregulations for both EMRT clusters and proteins tables ($P = 1$ means absolute upregulation and $P = 0$ means absolute downregulation).

References

1. MacNair JE, Lewis KC, Jorgenson JW (1997) Ultrahigh-pressure reversed-phase liquid chromatography in packed capillary columns. Anal Chem 69:983–989
2. MacNair JE, Patel KD, Jorgenson JW (1999) Ultrahigh-pressure reversed-phase capillary liquid chromatography: isocratic and gradient elution using columns packed with 1.0-μm particles. Anal Chem 71:700–708
3. Shen Y, Zhang R, Moore RJ, Kim J, Metz TO, Hixson KK, Zhao R, Livesay EA, Udseth HR, Smith RD (2005) Automated 20 kpsi RPLC-MS and MS/MS with chromatographic peak capacities of 1000–1500 and capabilities in proteomics and metabolomics. Anal Chem 77:3090–3100

4. Luo Q, Shen Y, Hixson KK, Zhao R, Yang F, Moore RJ, Mottaz HM, Smith RD (2005) Preparation of 20-microm-i.d. silica-based monolithic columns and their performance for proteomics analyses. Anal Chem 77:5028–5035
5. Xie C, Ye M, Jiang X, Jin W, Zou H (2006) Octadecylated silica monolith capillary column with integrated nanoelectrospray ionization emitter for highly efficient proteome analysis. Mol Cell Proteomics 5:454–461
6. Luo Q, Page JS, Tang K, Smith RD (2006) MicroSPE-nanoLC-ESI-MS/MS Using 10-μm-i.d. Silica-based monolithic columns for proteomics. Anal Chem 79:540–545
7. Luo Q, Tang K, Yang F, Elias A, Shen Y, Moore RJ, Zhao R, Hixson KK, Rossie SS, Smith RD (2006) More sensitive and quantitative proteomic measurements using very low flow rate porous silica monolithic LC columns with electrospray ionization-mass spectrometry. J Proteome Res 5:1091–1097
8. Miyamoto K, Hara T, Kobayashi H et al (2008) High-efficiency liquid chromatographic separation utilizing long monolithic silica capillary columns. Anal Chem 80:8741–8750
9. Zou H, Huang X, Ye M, Luo Q (2002) Monolithic stationary phases for liquid chromatography and capillary electrochromatography. J Chromatogr A 954:5–32
10. Wu R, Hu L, Wang F, Ye M, Zou H (2008) Recent development of monolithic stationary phases with emphasis on microscale chromatographic separation. J Chromatogr A 1184:369–392
11. Svec F (2004) Preparation and HPLC applications of rigid macroporous organic polymer monoliths. J Sep Sci 27:747–766
12. Svec F (2004) Organic polymer monoliths as stationary phases for capillary HPLC. J Sep Sci 27:1419–1430
13. Dong J, Zhou H, Wu R, Ye M, Zou H (2007) Specific capture of phosphopeptides by Zr4+-modified monolithic capillary column. J Sep Sci 30:2917–2923
14. Toll H, Wintringer R, Schweiger-Hufnagel U, Huber CG (2005) Comparing monolithic and microparticular capillary columns for the separation and analysis of peptide mixtures by liquid chromatography-mass spectrometry. J Sep Sci 28:1666–1674
15. Marcus K, Schäfer H, Klaus S, Bunse C, Swart R, Meyer HE (2006) A new fast method for nanoLC–MALDI-TOF/TOF–MS analysis using monolithic columns for peptide preconcentration and separation in proteomic studies. J Proteome Res 6:636–643
16. Mayr BM, Kohlbacher O, Reinert K, Sturm M, Gröpl C, Lange E, Klein C, Huber CG (2005) Absolute myoglobin quantitation in serum by combining two-dimensional liquid chromatography–electrospray ionization mass spectrometry and novel data analysis algorithms. J Proteome Res 5:414–421
17. Schley C, Altmeyer MO, Swart R, Müller R, Huber CG (2006) Proteome analysis of Myxococcus xanthus by off-line two-dimensional chromatographic separation using monolithic poly-(styrene-divinylbenzene) columns combined with ion-trap tandem mass spectrometry. J Proteome Res 5:2760–2768
18. Tholey A, Toll H, Huber CG (2005) Separation and detection of phosphorylated and nonphosphorylated peptides in liquid chromatography–mass spectrometry using monolithic columns and acidic or alkaline mobile phases. Anal Chem 77:4618–4625
19. Yue G, Luo Q, Zhang J, Wu S-L, Karger BL (2006) Ultratrace LC/MS proteomic analysis using 10-μm-i.d. Porous layer open tubular poly(styrene–divinylbenzene) capillary columns. Anal Chem 79:938–946
20. Luo Q, Yue G, Valaskovic GA, Gu Y, Wu S-L, Karger BL (2007) On-line 1D and 2D porous layer open tubular/LC-ESI-MS using 10-μm-i.d. poly(styrene–divinylbenzene) columns for ultrasensitive proteomic analysis. Anal Chem 79:6174–6181
21. Wu R, Zou H, Ye M, Lei Z, Ni J (2001) Capillary electrochromatography for separation of peptides driven with electrophoretic mobility on monolithic column. Anal Chem 73:4918–4923
22. Peters EC, Petro M, Svec F, Fréchet JMJ (1997) Molded rigid polymer monoliths as separation media for capillary electrochromatography. Anal Chem 69:3646–3649

23. Peters EC, Petro M, Svec F, Fréchet JMJ (1998) Molded rigid polymer monoliths as separation media for capillary electrochromatography. 1. Fine control of porous properties and surface chemistry. Anal Chem 70:2288–2295
24. Wang F, Dong J, Ye M, Wu R, Zou H (2009) Improvement of proteome coverage using hydrophobic monolithic columns in shotgun proteome analysis. J Chromatogr A 1216:3887–3894
25. Gu B, Chen Z, Thulin CD, Lee ML (2006) Efficient polymer monolith for strong cation-exchange capillary liquid chromatography of peptides. Anal Chem 78:3509–3518
26. Stanelle RD, Sander LC, Marcus RK (2005) Hydrodynamic flow in capillary-channel fiber columns for liquid chromatography. J Chromatogr A 1100:68–75
27. Eeltink S, Geiser L, Svec F, Frechet JM (2007) Optimization of the porous structure and polarity of polymethacrylate-based monolithic capillary columns for the LC-MS separation of enzymatic digests. J Sep Sci 30:2814–2820
28. Wang X, Barber WE, Carr PW (2006) A practical approach to maximizing peak capacity by using long columns packed with pellicular stationary phases for proteomic research. J Chromatogr A 1107:139–151
29. Wang X, Stoll DR, Schellinger AP, Carr PW (2006) Peak capacity optimization of peptide separations in reversed-phase gradient elution chromatography: fixed column format. Anal Chem 78:3406–3416
30. Link AJ, Eng J, Schieltz DM, Carmack E, Mize GJ, Morris DR, Garvik BM, Yates JR (1999) Direct analysis of protein complexes using mass spectrometry. Nat Biotech 17:676–682
31. Washburn MP, Wolters D, Yates JR (2001) Large-scale analysis of the yeast proteome by multidimensional protein identification technology. Nat Biotech 19:242–247
32. Wolters DA, Washburn MP, Yates JR (2001) An automated multidimensional protein identification technology for shotgun proteomics. Anal Chem 73:5683–5690
33. Wang F, Dong J, Ye M, Jiang X, Wu R, Zou H (2008) Online multidimensional separation with biphasic monolithic capillary column for shotgun proteome analysis. J Proteome Res 7:306–310
34. Motoyama A, Venable JD, Ruse CI, Yates JR (2006) Automated ultra-high-pressure multidimensional protein identification technology (UHP-MudPIT) for improved peptide identification of proteomic samples. Anal Chem 78:5109–5118
35. Ivanov AR, Zang L, Karger BL (2003) Low-attomole electrospray ionization MS and MS/MS analysis of protein tryptic digests using 20-microm-i.d. polystyrene-divinylbenzene monolithic capillary columns. Anal Chem 75:5306–5316
36. Wang F, Ye M, Dong J, Tian R, Hu L, Han G, Jiang X, Wu R, Zou H (2008) Improvement of performance in label-free quantitative proteome analysis with monolithic electrospray ionization emitter. J Sep Sci 31:2589–2597
37. Hughes MA, Silva JC, Geromanos SJ, Townsend CA (2006) Quantitative proteomic analysis of drug-induced changes in mycobacteria. J Proteome Res 5:54–63
38. Silva JC, Gorenstein MV, Li GZ, Vissers JP, Geromanos SJ (2006) Absolute quantification of proteins by LCMSE: a virtue of parallel MS acquisition. Mol Cell Proteomics 5:144–156
39. Silva JC, Denny R, Dorschel C, Gorenstein MV, Li GZ, Richardson K, Wall D, Geromanos SJ (2006) Simultaneous qualitative and quantitative analysis of the Escherichia coli proteome: a sweet tale. Mol Cell Proteomics 5:589–607
40. Vissers JP, Langridge JI, Aerts JM (2007) Analysis and quantification of diagnostic serum markers and protein signatures for Gaucher disease. Mol Cell Proteomics 6:755–766
41. Wang F, Dong J, Ye M, Wu R, Zou H (2009) Integration of monolithic frit into the particulate capillary (IMFPC) column in shotgun proteome analysis. Anal Chim Acta 652:324–330

Chapter 4
Large-Scale Proteome and Phosphoproteome Quantification by Using Dimethylation Isotope Labeling

4.1 Introduction

Protein characterization alone is usually not enough to elucidate most biological processes. Although thousands of proteins can be identified in one proteomic experiment, it is difficult to relate these proteins with their biological functions. The expression levels of proteins within a living organism are reflections of the different physiological and pathological states. Conventional protein quantification technologies such as western blot (WB) are low throughput and can only quantify one protein in each experiment. Therefore, comprehensive proteome quantification in certain depth is an important direction in the development of proteomic technologies and is considered as the bridge for the gap between proteins and their biological function [1–7]. On the other hand, there are more than 300 types of posttranslational modifications (PTMs) that can dynamically modify the whole proteome, which makes the protein sample even more complex. The occupancy level of a PTM on the specific site of a protein is also critical to the biological function of the protein in the regulation of different physiological and pathological processes, such as the protein phosphorylation is related to the signal transduction in many pathways activation and protein glycosylation is related to cell-to-cell recognition [8–10]. Therefore, comprehensive quantification of proteome PTMs is also another important task for current proteomic analysis.

Two strategies are feasibly developed for the comprehensive proteome quantification in mass spectrometry (MS)-based shotgun proteomics. The first is the label-free approach, which obtains the relative quantity of each peptide among different samples by comparing the corresponding peak intensity in parallel nanoflow liquid chromatography coupled with tandem mass spectrometry (LC-MS/MS) analyses [11, 12]. The advantage of the label-free approach is that no isotope labeling is required, and multiple protein samples can be compared simultaneously. However, the quantification accuracy largely depends on the reproducibility of nanoflow LC–MS/MS analysis. The other is the stable isotope labeling strategy, which introduces different mass differentiate isotope tags into different samples at first, and

F. Wang, *Applications of Monolithic Column and Isotope Dimethylation Labeling in Shotgun Proteome Analysis*, Springer Theses, DOI: 10.1007/978-3-642-42008-5_4,

then the samples are mixed together and analyzed by LC-MS/MS simultaneously. The relative quantity of each peptide among different samples is obtained by comparing the intensity of specific isotopic peaks in the corresponding MS spectrum [13–15]. Stable isotope labeling with amino acids in cell culture (SILAC) introduces isotope tags by in vivo cell culture, which can compare the expression levels of thousands of proteins in a single labeling experiment using data analysis software MSQuant or MaxQuant [16–18]. Recently, the proteins in living mouse could be also isotope labeled by dosing a diet containing either the natural or the $^{13}C_6$ substituted version of lysine [6]. However, SILAC strategy needs expensive reagents and it is impossible to obtain SILAC clinical samples of humans, such as serum, urine, and tissues. Dimethyl isotope labeling at peptide level is another popular in vitro isotope labeling strategy for proteome quantification, which globally labels the N-terminus and ε-amino group of lysine with dimethyl groups using inexpensive isotopic formaldehyde and sodium cyanoborohydride [19–21]. Conventional isotope labeling procedures usually need many manually handling steps such as sample desalting, adding labeling regents, incubation, quenching labeling, sample redesalting, and lyophilization, which may result in sample loss and introduce unexpected contaminants [22]. Thus both the quantification reproducibility and accuracy are dependent on the high quality sample preparation. Moreover, the manually handling procedure is time-consuming and labor-intensive. To circumvent these limitations, Raijmakers et al. established an online sequential isotope dimethyl labeling method, which allows sample loading, isotope dimethyl labeling, and one-dimensional (1D) reversed phase (RP) LC–MS/MS analysis in a fully automated manner [23].

Hepatocellular carcinoma (HCC) is the most common primary cancer of the liver, especially in Africa, Southeast Asia, and China [24, 25]. Stable isotope labeling strategy was also applied to quantify the aberrant protein expression level and PTMs of HCC samples. Chaerkady et al. quantified over 600 proteins among which 59 and 92 proteins were found up- and downregulated, respectively, in HCC liver tissues by using isobaric tagging for relative and absolute quantification (iTRAQ) technique [26]. Chen et al. quantified 2,335 proteins and found 91 and 61 proteins were up- and down-regulated in HCC cell line HCCLM6 as compared to MHCC97L [27]. However, all of these isotope labeling approaches were manually performed and offline SCX chromatography or SDS-PAGE was used for peptides or proteins prefractionation. Further, the phosphoproteome of HCC liver tissue is also not investigated enough, which greatly limits our understanding of the biological function of phosphorylation in the process of HCC development.

Alzheimer's disease (AD) affects 35 million people worldwide. AD is the most prevalent form of old age dementia in humans. AD neuropathology is defined as: (1) intraneuronal neurofibrillary tangles composed of hyperphosphorylated tau and (2) aberrant processing of the amyloid precursor protein (APP) to Aβ. The "amyloid cascade hypothesis" defines proteolytic cleavage of APP into smaller, toxic Aβ fragments as the root cause of AD [28, 29]. However, in the early onset form of AD, which exhibits different mutations in the amyloid precursor protein (APP), the molecular process linked to Aβ metabolism is not clear. Therefore, various transgenic mouse models were established to study the effects of specific genes involved in an early onset of AD [30–33]. Chishti et al. created the TgCRND8 mouse model, which has a double mutant form of amyloid precursor protein 695 (APP695) with both the Swedish and Indiana mutations and exhibits Aβ-amyloid plaque deposition and robust cognitive deficits from the age of 3 months [34]. This transgenic mouse model is widely used in studying the molecular mechanism and for large-scale "omics studies" of AD and the formation of Aβ-amyloid plaques [35–37]. However, there has not been any systematic proteomic study of the role of protein phosphorylation during the pathological development of the TgCRND8 model.

In this chapter, we first combined the online isotope labeling with online nanoflow two dimensional (2D) LC-MS/MS analysis. All the procedures for isotope dimethyl labeling as well as SCX-RP 2D LC-MS/MS analysis were automatically performed. Therefore, no manual operation is needed for this online proteome quantification system, which greatly reduces the sample loss and unexpected contaminants. Then, this online proteome quantification system was applied to quantify the different proteome expression of HCC and normal liver tissues, and totally 554 proteins were reliably quantified and 94 proteins exhibited upregulated and 249 proteins exhibited downregulated. In order to further improve the proteome quantification accuracy, a pseudo triplex isotope labeling strategy was developed. Briefly, two identical samples are labeled with light and heavy isotopes, respectively, while another comparative sample is labeled with a medium isotope. Two replicated quantification results were achieved in just one experiment, and the coefficient of variation (CV) criterion was used to control the quantification accuracy. Finally, this pseudo triplex isotope labeling strategy was successfully applied to investigate the different phosphoproteome expressions of HCC and normal liver tissues as well as hippocampus phosphoproteome changes in 2- and 6-month-old TgCRND8 mice and congenic littermate controls.

4.2 Results

4.2.1 Online Isotope Labeling with Nanoflow 2D LC-MS/MS Analysis

Conventional isotope labeling strategies need tedious manual operations, such as desalting, labeling, purification, and lyophilization, and variations and contaminants are easily introduced during each of these steps, which will decrease the accuracy and throughput of proteome quantification significantly. Raijmakers et al. first developed online sequential isotope dimethyl labeling, followed with 1D RP LC-MS/MS analysis [23]. However, the separation capability of 1D RP gradient LC separation is not always enough when complex biological samples are analyzed. In order to increase the proteome coverage in large-scale protein

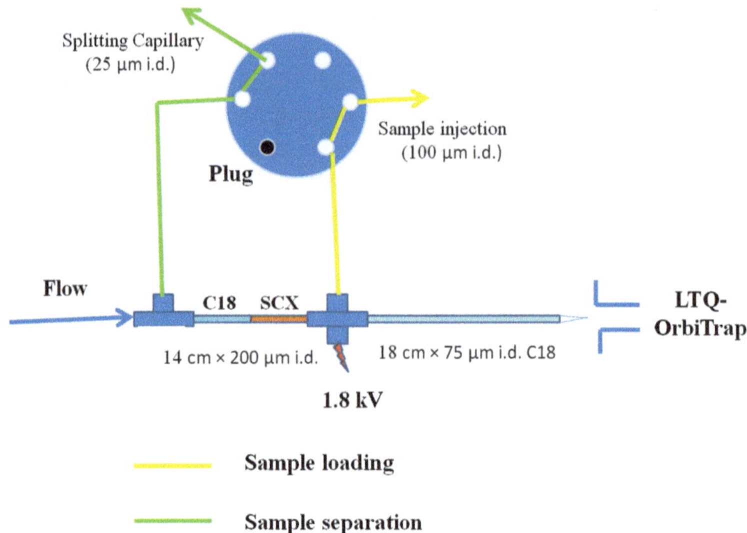

Fig. 4.1 Schematic diagram of the vent sample injection system using C18-SCX biphasic column for online isotope labeling and online 2D LC separation (Redrawn with permission of Ref. [38]. Copyright 2010 American Chemical Society)

quantification, especially to quantify proteins with relatively low abundance, it is necessary to increase the separation capability of the LC system. SCX- RP online multidimensional separation exhibits high separation capability and advantages such as higher sensitivity, no manual interference, and so on. Thus, coupling online isotope labeling with online multidimensional separation can further increase the proteome quantification sensitivity, reproducibility, and accuracy. However, as the pH value of the online isotope dimethyl labeling regents is optimized at 7.5, it is improper to perform isotope labeling of peptides onto SCX trap column. Therefore, an RP-SCX biphasic trap column was prepared and applied in the vented sample injection system as shown in Fig. 4.1. The RP segment of the biphasic column was used for isotope dimethyl labeling, and the labeled peptides were transferred to the SCX segment by ACN elution. Then, online SCX-RP 2D LC separation was performed by stepwise salt elution of the SCX trapped peptides to the C18 separation column as described in the previous chapters. As the phosphate SCX monolith has superior permeability and sample loading capacity as described in Chap. 2, high-resolution online salt stepwise fractionation can be realized to increase the separation capability of the LC system efficiently.

To evaluate the performance of the system for automatically isotope labeling and online 2D LC-MS/MS analysis, the same amount (15 μg) tryptic digests of extracted protein from normal human liver tissues were injected in sequence as testing sample and labeled with light and heavy dimethyl groups, respectively. Then, 29-h online 2D LC separation with 11 fractions was performed. Finally 30,738 peptides (corresponding to 1,115 proteins) with Mascot score $\geq$25

Table 4.1 The number of quantified proteins and peptides by each analysis

	Control-1	Control-2	Control-3	29 h-1	29 h-2	29 h-3	63-h
Pro. num.	954	823	1,023	1,181	1,096	1,119	1,767
Pep. num.	4,425	3,855	5,303	5,473	4,826	4,982	12,129
FDR (Pro.) (%)	2.20	1.58	1.86	1.61	0.73	1.61	1.19
FDR (Pep.) (%)	0.54	0.36	0.38	0.38	0.16	0.38	0.20

(rank 1, $P \leq 0.05$) were obtained at FDR 1.2 % and 29,483 peptides (95.9 %) were successfully labeled with dimethyl groups. In all the labeled peptides, 320 peptides only have dimethylated lysine, 15,668 peptides only have dimethylated N termini, and 13,495 peptides have both dimethylated lysine and N termini. Among the 15,668 peptides that only have the dimethylated N termini, 11,821 peptides do not have lysine residue, and over 85.0 % ((11,821 + 13,495)/29,483) of the labeled peptides were completely labeled. The numbers of identified peptides with light and heavy dimethyl labels are 15,135, and 14,348, respectively, and the ratio is 1.05. Therefore, little labeling interference was observed between the sequentially injected and labeled samples. Then, these identified peptides were paired with the MS spectra using MSQuant and the quantification results were combined and normalized by StatQuant as described in experimental section. Finally, 4,425 peptides corresponding to 954 proteins were successfully quantified (Control 1, Table 4.1), and the identification FDR were 2.20 and 0.54 % for proteins and peptides, respectively. A total of 597 proteins (62.6 %) were quantified by at least two peptides. The typical isotope isoforms of peptides R.MFLSFPTTK.T and K.TCNVLVALEQQSPDIAQCVHLDR.N in MS spectra were given in Fig. 4.2a, b. Obviously, the peptides were reliably labeled with dimethyl groups and could be quantified by comparing the intensity of the isotope isoforms. The log2 ratios of the quantified proteins were shown in Fig. 4.3a. Though most of the log2 ratios were around 0 (ratio 1), some of the values were widely scattered between −6 and 6 and the average value of the log2 ratios is 0.220 (theoretically 0), which is not acceptable. Among the 179 proteins who have changed their quantity twofolds (theoretically 1), 93 proteins (52.0 %) were quantified by at least two peptides. Therefore, if we just used the filtering criterion that proteins must be quantified by at least two peptides, 52.0 % false quantification can pass the filter criterion but accurate quantification of many proteins by one peptide is filtered. Therefore, it is necessary to establish a more suitable filtering criterion in isotope labeling quantitative proteome analysis.

Then, two replicated online isotope dimethyl labeling and 2D LC separation as described above were performed and finally 3,855 and 5,303 peptides and 823 and 1,023 proteins were successfully quantified, respectively (control-2 and control-3, Table 4.1). Totally 588 proteins were quantified in all of the three replicated analyses and the relative quantity of 394 proteins had relative standard deviation (RSD) <50 %. The log2 ratios (average value of the three analyses) of the 394 proteins are shown in Fig. 4.3b, and it can be seen that there is no protein with log2 ratio > 1

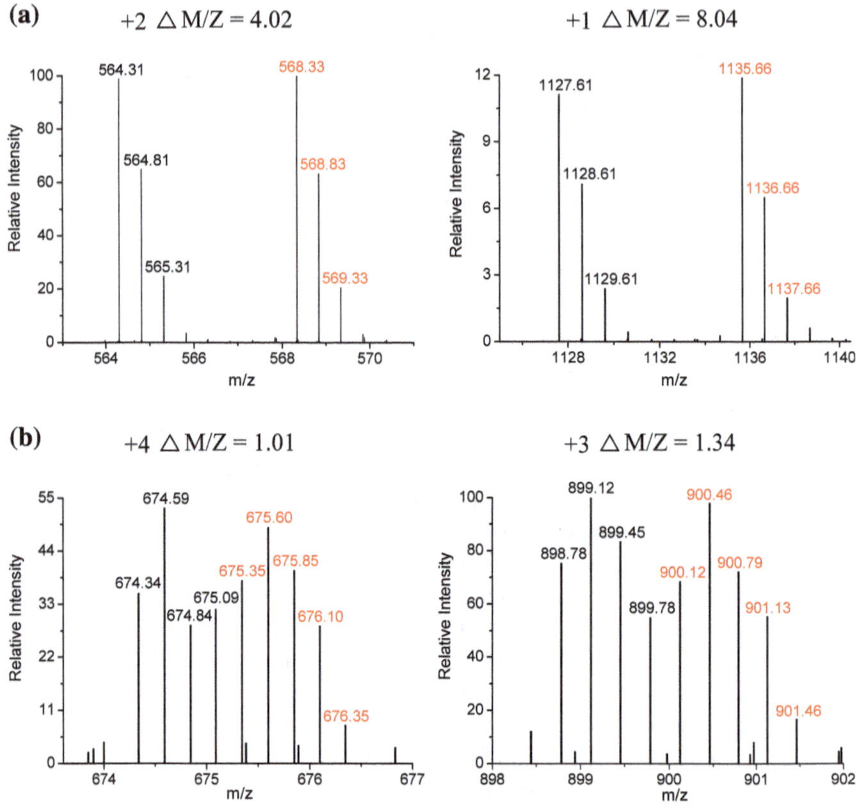

Fig. 4.2 The peaks for isotope isoforms of peptides **a** R.MFLSFPTTK.T (singly and doubly charged) and **b** K.TCNVLVALEQQSPDIAQCVHLDR.N (doubly and triply charged) in MS spectra (Reprinted with permission of Ref. [38]. Copyright 2010 American Chemical Society)

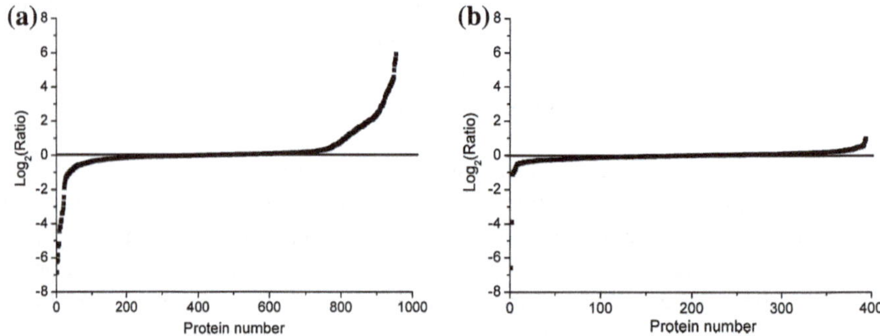

Fig. 4.3 Protein expression ratios (Normal/Normal) in log2 obtained by **a** the sequential light/heavy isotope labeling coupled with 29-h online multidimensional separation; and **b** three times replicated quantification analyses with RSD < 50 % (Reprinted with permission of Ref. [38]. Copyright 2010 American Chemical Society)

and only 4 protein with log2 ratio < −1. Therefore, only 1.0 % (4) of the quantified proteins in replicated analyses has changed their quantity more than twofolds (theoretically 1). And the average value of the log2 ratios is −0.05, which is much better than the result obtained by just one time analysis. In the other 194 proteins that quantified in all of the three replicated analyses but have RSD > 50 %, only two proteins (1.0 %) have simultaneously changed relative quantity more than twofolds. Therefore, these results indicate that the quantification accuracy is significantly improved by three times replicated analyses and control the reproducibility. Similar results were also obtained in online sequential isotope dimethyl labeling combined with 1D analysis (data not shown). Then three replicated analyses were conducted in the quantitative proteome analysis of the HCC and normal liver tissues lately. And the proteins quantified in all of the three analyses with RSD < 50 % were considered as reliable quantification. The proteins with average log2 ratio >1 or <−1 were considered as significantly upregulated or downregulated. Furthermore, the proteins exhibited, simultaneously upregulated (log2 ratio > 1) and downregulated (log2 ratio < 1), in all of the three replicated analyses were also considered as significantly regulated proteins even if their RSD > 50 %.

4.2.2 Comparative Proteome Quantification of HCC and Normal Liver Tissues by Online Isotope Labeling

Tryptic digests of the same amount of proteins (15 μg) extracted from HCC and normal human liver tissues were injected in sequence and labeled with light and heavy isotope dimethyl groups, respectively. Then, the 29-h online 2D LC separation coupled with MS/MS detection was performed and the LC-MS/MS base peak chromatograms for the 11 fractions as shown in Fig. 4.4. The peptides are symmetrically distributed among the fractions and finally 1,181 proteins could be quantified (29 h-1, Table 4.1). Then, the whole procedures for sequential isotope dimethyl labeling and online 2D LC-MS/MS analysis were repeated twice, and 1,096 and 1,119 proteins were successfully quantified (29 h-2 and 29 h-3, Table 4.1). Therefore, the average numbers of quantified peptides and proteins are 5,094 (RSD = 6.6 %, $n = 3$) and 1,132 (RSD = 3.9 %, $n = 3$). Within these three replicated analyses, 797 proteins were quantified in all of the three analyses. The relative quantity of 483 proteins have RSDs < 50 %. Among the proteins with RSD > 50 %, 71 proteins exhibit simultaneously upregulation (ratio > 2) or downregulation (ratio < 0.5) in the replicated analyses, and we believe these proteins are also reliably quantified as described above. Therefore, totally 554 proteins are reliably quantified and 94 proteins exhibit upregulation and 249 proteins exhibit downregulaion.

Different expressions of 218 genes were detected with cDNA microarrays in HCC tumor compared to nontumor liver samples by Neo et al. [39]. Among the 554 reliably quantified proteins in our study, the genes of 21 proteins have also been studied by Neo et al. as summarized in Table 4.2. Among the 21 proteins,

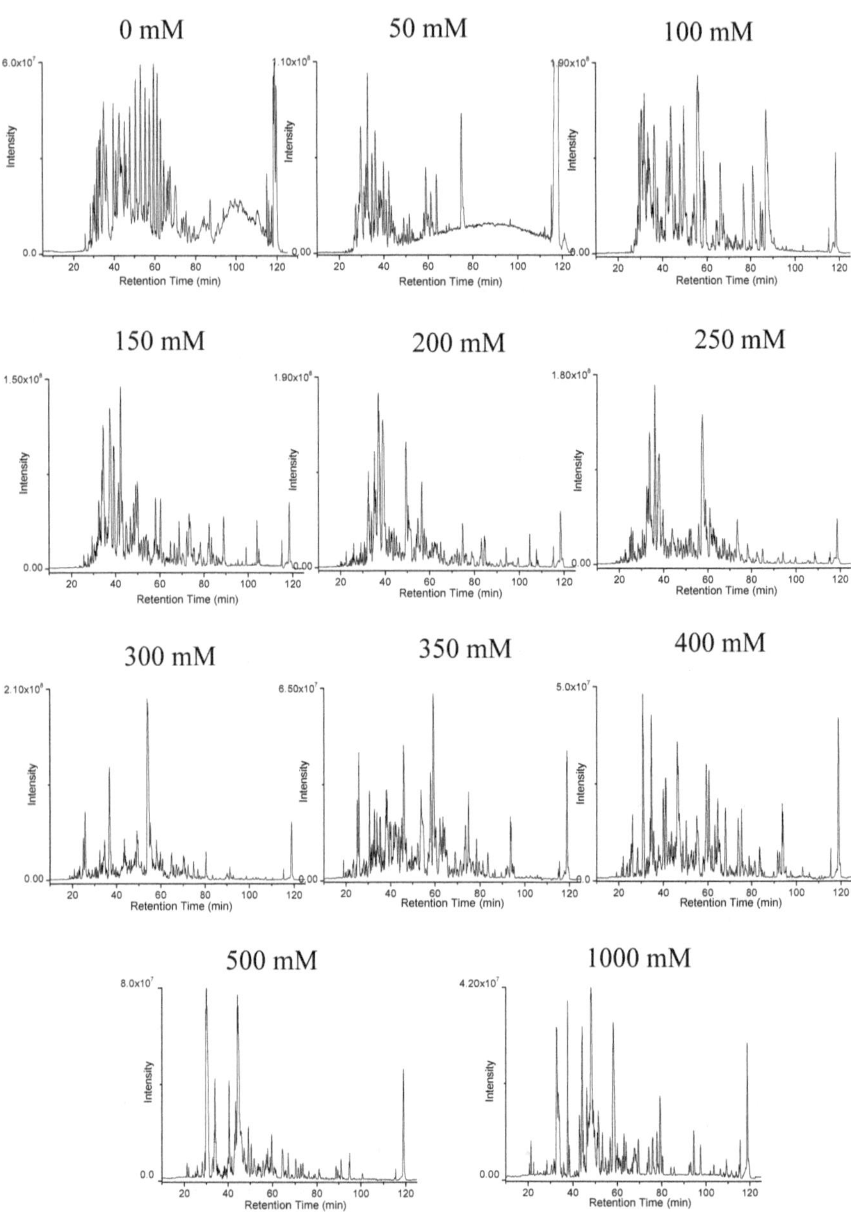

Fig. 4.4 Base peak chromatograms of an 11-cycle online 2D LC-MS/MS analysis of isotope dimethyl labeled samples (HCC liver sample labeled with light dimethyl groups and normal liver sample labeled with heavy dimethyl groups by the online isotope labeling strategy) (Reprinted with permission of Ref. [38]. Copyright 2010 American Chemical Society)

Table 4.2 List of the quantified proteins with the genes expression level studied (HCC/Normal) by Neo et al. [39]

No.	IPI No.	Gene symbol and protein name	Ratio	Ref. [38]
1	IPI:IPI00027626.3	CCT6A T-complex protein 1 subunit zeta	10.11	↑
2	IPI:IPI00013415.1	RPS7 40S ribosomal protein S7	7.92	↑
3	IPI:IPI00216318.5	YWHAB isoform long of 14-3-3 protein beta/alpha	4.69	↑
4	IPI:IPI00021290.5	ACLY ATP-citrate synthase	4.42	↑
5	IPI:IPI00418169.3	ANXA2 annexin A2 isoform 1	2.54	↑
6	IPI:IPI00444262.3	NCL cDNA FLJ45706 fis, clone FEBRA2028457, highly similar to Nucleolin	1.58	↑
7	IPI:IPI00003865.1	HSPA8 isoform 1 of heat shock cognate 71 kDa protein	1.44	↑
8	IPI:IPI00010740.1	SFPQ isoform long of splicing factor, proline- and glutamine-rich	1.39	↑
9	IPI:IPI00026182.5	CAPZA2 F-actin-capping protein subunit alpha-2	1.17	↑
10	IPI:IPI00003362.2	HSPA5 HSPA5 protein	1.16	↑
11	IPI:IPI00215884.4	SFRS1 isoform ASF-1 of splicing factor, arginine/serine-rich 1	1.15	↑
12	IPI:IPI00022774.3	VCP transitional endoplasmic reticulum ATPase	1.05	↑
13	IPI:IPI00419585.9	PPIA peptidyl-prolyl cis-trans isomerase A	0.84	↑
14	IPI:IPI00296337.2	PRKDC isoform 1 of DNA-dependent protein kinase catalytic subunit	0.84	↑
15	IPI:IPI00030131.3	TMPO isoform beta of lamina-associated polypeptide 2, isoforms beta/gamma	0.66	↑
16	IPI:IPI00010720.1	CCT5 T-complex protein 1 subunit epsilon	0.65	↑
17	IPI:IPI00290770.3	CCT3 chaperonin containing TCP1, subunit 3 isoform b	0.63	↑
18	IPI:IPI00023640.3	PDCD5 programmed cell death protein 5	0.46	↑
19	IPI:IPI00012828.3	ACAA1 3-ketoacyl-CoA thiolase, peroxisomal	0.14	↓
20	IPI:IPI00473031.6	ADH1B alcohol dehydrogenase 1B	0.03	↓
21	IPI:IPI00290301.1	CYP2C8 cytochrome P450 2C8	0.10	↓

↑: gene expression upregulated; ↓: gene expression downregulated

T-complex protein 1 subunit zeta (CCT6A), 40S ribosomal protein S7 (RPS7), isoform long of 14-3-3 protein beta/alpha (YWHAB), ATP-citrate synthase (ACLY), and annexin A2 isoform 1 (ANXA2) are considered as upregulated; 3-ketoacyl-CoA thiolase peroxisomal (ACAA1), cytochrome P450 2C8 (CYP2C8), and alcohol dehydrogenase 1B (ADH1B) are determined to be downregulated in our isotope dimethyl labeling quantification, which are consistent with the status of genes expression. The only exception is programmed cell death protein 5 (PDCD5), which is downregulated in protein quantification (ratio 0.46), but its gene expression shows upregulated. Though the relative quantity of the other 12 proteins is considered as nonchanged in our study, many of them also exhibit upregulation trends as summarized in Table 4.2. Therefore, most of the gene regulation level (~70 %) is consistent with our protein quantification results.

Chaerkady et al. quantified over 600 proteins (HCC/Normal) using the iTRAQ technique [26]. Among these proteins, 263 proteins were also quantified in our experiment. 11 proteins are upregulated and 60 proteins are downregulated simultaneously in the both quantification results. Among these proteins, extracellular superoxide dismutase [Cu-Zn] (SOD3), Serotransferrin (TF), and isoform long of 14-3-3 protein beta/alpha (YWHAB) are also reported as upregulated proteins in other studies [39–41]. Argininosuccinate lyase (ASL), argininosuccinate synthase (ASS1), ornithine carbamoyl transferase (OTC), and carbamoyl phosphate synthase (CPS1) are all enzymes involved in urea cycle, which is an essential metabolic pathway of liver for detoxification of ammonia. And they are all significantly downregulated in the HCC tissues. Aspartate amino transferase (GOT), which is involved in intermediary metabolism, associate with urea cycle, is also downregulated. Chaerkady et al. also further validated the downregulation of CPS1 and OTC in HCC samples by WB and gene expression studies [26].

Although many technologies have been developed to study the protein expression level of clinical samples, a strategy with high throughput and reproducibility is still needed for large-scale proteins quantification. In the above two sections, a system with automatically sequential isotope dimethyl labeling coupled with online multidimensional separation was developed for protein quantification by using an RP-SCX biphasic trap column. In this system, all procedures including sample injection, sequential sample labeling, online stepwise fractionation, and nanoflow LC-MS/MS analysis are automatically operated, which is beneficial for improving the reproducibility and accuracy of quantification. Furthermore, the proteome coverage can be efficiently increased along with the increase in the number of stepwise fractions and the time of RP separation gradient. In our experiments, we also increased the number of salt step fractions and extended the reversed phase binary gradient to develop a 63-h online 2D LC-MS/MS analysis. Finally, the relative quantification information about 12,129 peptides corresponding to 1,767 proteins was obtained (63-h, Table 4.1). Therefore, as the analysis time for online 2D LC-MS/MS analysis increased to 117.2 %, the number of quantified peptides and proteins increased to 138.1 and 56.1 %, respectively. It can be seen that increasing the number of stepwise fractions and gradient separation time is a very effective method to increasing the proteome coverage in large-scale proteome quantification. Another advantage of this system is that a 7-cm-long phosphate SCX monolith is integrated into the biphasic trap column. This type of phosphate SCX monolith has much higher permeability and sample loading capacity compared to commercially available SCX materials, which is beneficial for improving the resolution in SCX fractionation as described in Chap. 2. Although 14-cm-long biphasic trap column was utilized, sample loading and isotope labeling could be operated at back pressure less than 1,000 psi. By comparing with the quantification of about 1,400 proteins by applying 25 offline SCX fractionation for tryptic digest of about 180 μg proteins extracted from zebrafish embryos [42], or about 600 proteins by using 39 offline SCX fractionation for tryptic digest of about 160 μg proteins from human livers [26], about 1,000 proteins were quantified by applying 11 online SCX fractionation for tryptic digest of about 30 μg proteins from human liver by our system, which exhibits better proteome coverage and detection sensitivity.

4.2.3 Pseudo Triplex Isotope Labeling Strategy

As described in Sect. 4.1, the proteome quantification accuracy is not acceptable if only one-time quantitative analysis is performed. The relative ratios of about 10 % of the quantified peptides are <0.5 or >2 even if the same amounts of comparative samples are utilized. We further investigated the relationship of quantification accuracy and reproducibility. Three identical protein digests of 50 μg mouse brain protein were labeled with light (L), medium (M), and heavy (H) dimethyl group, respectively. Then, the three samples were mixed together and analyzed by 2D LC-MS/MS, and two-time replicates were performed. 950 distinct protein groups could be successfully quantified from 3,500 unique peptides in one-time quantification. Finally, 2,162 and 2,231 unique peptides could give the relative ratio of M/L and H/L, respectively, in both replicate proteome quantification. Then, log2 ratios (average value) were plotted with corresponding CV in the two replicate analyses (Fig. 4.5). Obviously, the quantification error is increased along with the increase in CV, and most of the peptides with relative ratios changed more than two times located in the range of CV > 40 %. Therefore, we set CV < 40 % as a filtering criterion to filter most of the bad quantified results. About 600 proteins could be quantified from about 2,000 unique peptides in this criterion, and 5 (0.8 %) and 16 (3 %) proteins exhibited relative ratios M/L and H/L changed more than two times. If three times replicate quantification were performed and also controlled the CV < 40 % for peptides quantified in all of the three times, the quantification accuracy could be further improved. However, only about 400 proteins were obtained at this condition, which decreased about 60 % compared with just one-time quantification. Therefore, it is still a challenge to improve the proteome quantification accuracy without decrease in the proteome coverage.

In conventional triplex isotope labeling strategy, three different protein samples are labeled with light, medium, and heavy isotope regents, and the relative ratio among these three different protein samples can be obtained (Fig. 4.6a). Replicate quantitative proteome analysis needs to be performed to improve the quantification

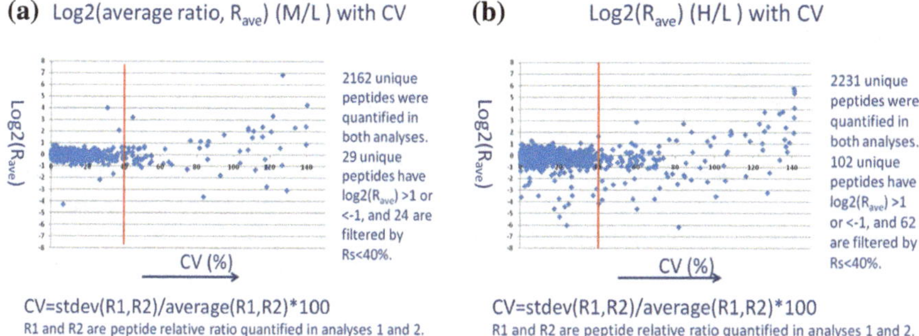

Fig. 4.5 The quantification error increases along with the CV increase for both **a** intermediate/light and **b** heavy/light in replicate analyses

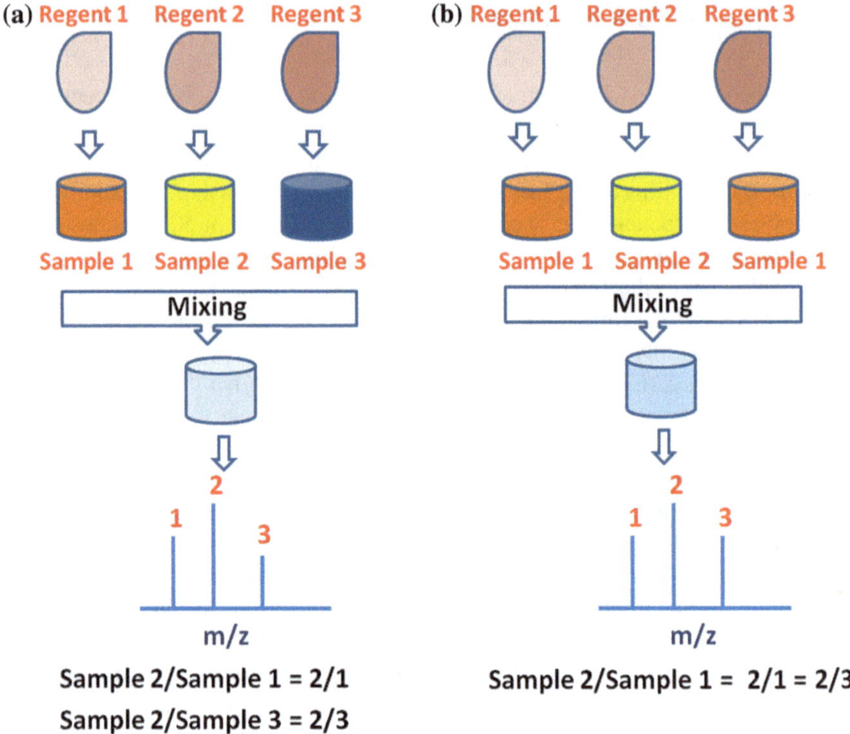

Fig. 4.6 Schematic diagrams for **a** common labeling strategy and **b** replicate labeling strategy used in this study

accuracy as described above. However, the overlap of quantified proteins is only about 60–80 % and 40–60 % for two and three times replicate analyses, respectively, due to the randomness of data-dependent analysis (DDA) mode in shotgun proteomics. Thus, a pseudo triplex isotope labeling strategy was developed. Briefly, two identical control samples are labeled with light and heavy isotopes, respectively, while another comparative sample is labeled with a medium isotope (Fig. 4.6b). Thus, two replicate analyses can be achieved in just one LC-MS/MS experiment as the M/L and H/L both reveal the relative ratio between comparative and control samples. We plot the average values of log2 ratio of the 3,405 quantified unique peptides with the CV between M/L and M/H as shown in Fig. 4.7a. Similar to conventional replicate proteome quantification (Fig. 4.5a), the proteome quantification error increased along with the CV. And about 90 % of the peptides with relative ratio <0.5 or >2 could be filtered by CV < 40 %, which is similar to three times conventional replicate analyses. Finally, 819 distinct protein groups were quantified, among which only five proteins (<1 %) with relative ratio changed more than two times. Therefore, the number of quantified proteins increased about 100 % compared with convention three times replicate analyses

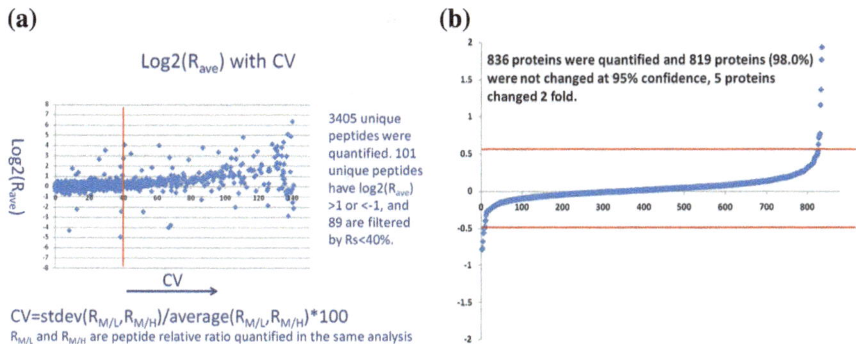

Fig. 4.7 **a** The quantification error increases along with the CV increase in the same replicate labeling quantification experiment; **b** proteins log2 ratios distribution by one time replicate labeling quantification analysis and CV < 40 % filtering (the sample labeling amounts ratio is L:M:H = 1:1:1)

and controlling the two times changed proteins <1 %. To calculate the 95 % confidence interval (median (all ratios) $\pm$ 1.96 $\times$ stdev (all ratios)), 98 % of the quantified proteins located in the no changed interval. And the average 95 % confidence interval for three times pseudo triplex isotope labeling proteome quantification is [0.7–1.5]. Therefore, if the protein with relative ratio is out of this interval, we can consider it as significant changed protein (95 % confidence) in real sample analysis. Then, the three triplex isotope labeled protein samples were mixed with a ratio of 1:4:1 to get a more complex testing sample for the pseudo triplex isotope labeling strategy. Finally, 4,333 unique peptides corresponding to 866 distinct proteins were successfully quantified and similar quantification accuracy were also obtained (Fig. 4.8a, b), and 96 % of the quantified proteins located in the no changed interval at 95 % confidence.

We also investigated if this pseudo triplex isotope labeling strategy can be utilized for phosphoproteome quantification. Three identical protein samples (200 μg each) were labeled with light, medium, and heavy isotope dimethyl group, respectively. Then, these three protein samples were mixed together, followed with phosphopeptides enrichment by Ti^{4+}-IMAC. After 2D LC-MS/MS analysis, 884 unique phosphopeptides were quantified. The average values of log2 ratios (M/L and M/H) were also plot against the CV (M/L and M/H) as shown in Fig. 4.9a. Similar to the results of proteome quantification, the quantification errors were also increased along with the CV. The relative ratios of only 13 unique phosphopeptides were changed more than two times, which demonstrated that better quantification accuracy can be achieved for phosphoproteome analysis. This is because the enrichment of phosphopeptides by Ti^{4+}-IMAC decreases the complexity of the sample. After controlling CV < 40 %, 828 phosphopeptides were accurately quantified and no phosphopeptide changed the ratio more than two times.

About 10 % of the quantified peptides change the relative ratio more than two times in one-time quantitative proteome analysis even if identical protein sample

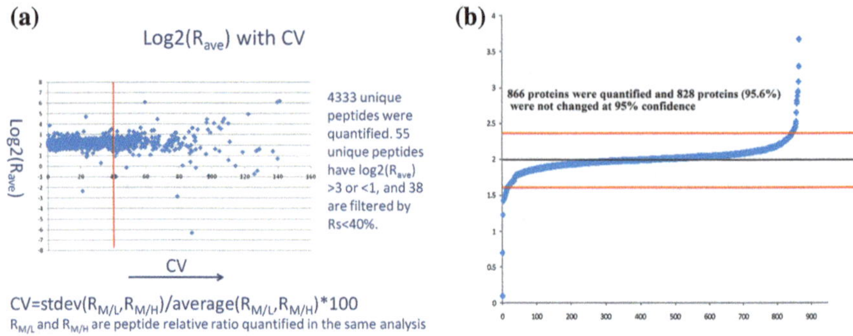

Fig. 4.8 **a** The quantification error increases along with the CV increase in the same replicate labeling quantification experiment; **b** proteins log2 ratios distribution by one-time replicate labeling quantification analysis and CV < 40 % filtering (the sample labeling amounts ratio is L:M:H = 1:4:1)

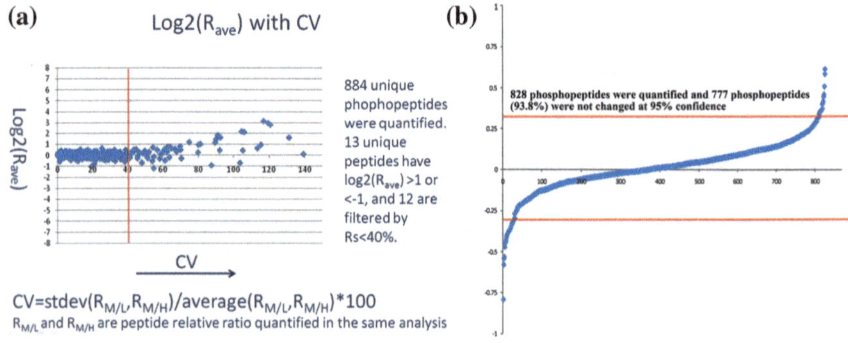

Fig. 4.9 **a** The phosphopeptides quantification error increases along with the CV increase in the same replicate labeling quantification experiment; **b** phosphopeptides log2 ratios distribution by one-time replicate labeling quantification analysis and CV < 40 % filtering (the sample labeling amounts ratio is L:M:H = 1:1:1, and 200 μg of each labeled sample was used for phosphopeptides enrichment)

are utilized as described above. Therefore, replicate quantitative analyses are needed to improve the quantification accuracy. However, the proteome overlap among replicate analyses significantly decrease along with the times of replication due to the randomness of DDA MS detection modes in shotgun proteomics. It is a challenge to get quantification results with both high proteome coverage and high accuracy. In this section, we developed a pseudo triplex isotope labeling strategy, which can give two times replicate proteome quantification in just one experiment. Comparing to conventional replicate proteome quantification, only half amount of the comparative sample is needed, and the number of quantified proteins increases 100 % at identical quantification accuracy. Further, this pseudo triplex isotope labeling strategy can be also successfully applied for high accurate phosphoproteome quantification. Therefore, we think this strategy has great potential in comprehensive proteome and proteome PTMs quantification.

4.2.4 Application of the Pseudo Triplex Labeling Approach to Comparative Phosphoproteome Quantification of Human Liver Tissues

Phosphorylation is an important protein posttranslational modification (PTM) in diverse biological processes, such as signal transduction and cell cycle. The changes in signaling occurring in HCC is reflected in part by change in protein phosphorylations. Although a few proteomic/phosphoproteomic studies of HCC have been reported [43, 44], the coverage of the HCC phosphoproteome is very limited. Therefore, we applied the pseudo triplex dimethyl labeling approach for the comprehensive quantitative phosphoproteome analysis of HCC and normal human liver tissues.

Briefly, two tryptic digests of 1 mg of protein extracted from normal human liver tissue were used as controls and labeled with the light (L) and heavy (H) dimethyl groups, respectively. A third tryptic digests of 1 mg protein extracted from HCC sample was labeled with the medium (M) dimethyl groups. Then the triplex labeled samples were mixed together at a ratio of 1:1:1 and enriched by the Ti^{4+}-IMAC. To increase the phosphoproteome coverage, a 42-h online 2D LC separation with 15 salt gradients was performed. Finally, 1,934 phosphopeptides corresponding to 1,033 phosphoproteins were successfully quantified in the HCC and normal human liver tissues. After filtering the dataset using CV $<$ 50 % (M/L and M/H), a total of 1,831 phosphopeptides corresponding to 1,918 phosphorylation sites were reliably quantified. A conservative threshold of twofold change was utilized for the selection of significantly changed phosphopeptides. Finally, 471 phosphopeptides (478 phosphorylation sites) were downregulated whereas 309 phosphopeptides (332 phosphorylation sites) were upregulated in the HCC versus normal human liver tissues (Fig. 4.10). Biological function analyses of the down- and up-regulated phosphoproteins were performed as shown in Fig. 4.11.

The quantitative phosphoproteome analysis of human liver tissues was performed by Lee et al. using the $H_2{}^{18}O$-labeled approach combined with a LTQ mass spectrometer [44]. Two potential HCC phosphor-biomarkers (plectin-1, phospho-Ser-4,280 and alpha-HS-glycoprotein, phospho-Ser 138) were quantified in their study. Interestingly, both of these phosphopeptides were quantified in our study and the quantification results were consistent. Moreover, over 70 % of the quantified phosphorylation sites in our study have been reported in other qualitative or quantitative phosphorylation analyses. All of these observations demonstrate the high confidence of this comprehensive quantitative phosphorylation analysis of human liver. To the best of our knowledge, this is the largest quantitative phosphoproteome dataset from the human liver.

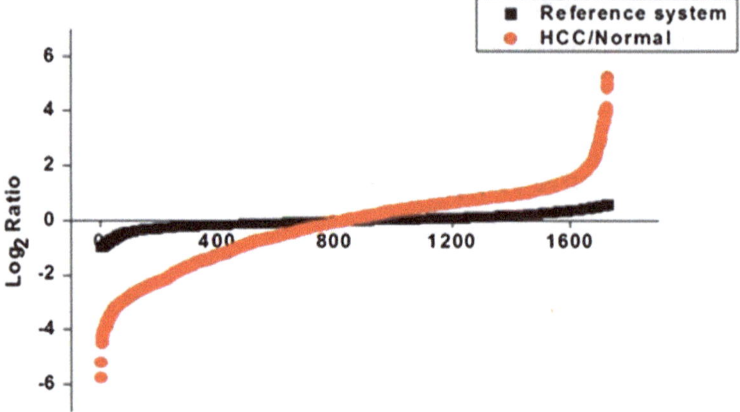

Fig. 4.10 The log2 ratio distributions of the quantified phosphopeptides in normal to normal (reference system) and HCC to normal human liver tissues (Reprinted with permission from Ref. [45]. Copyright 2011 American Chemical Society)

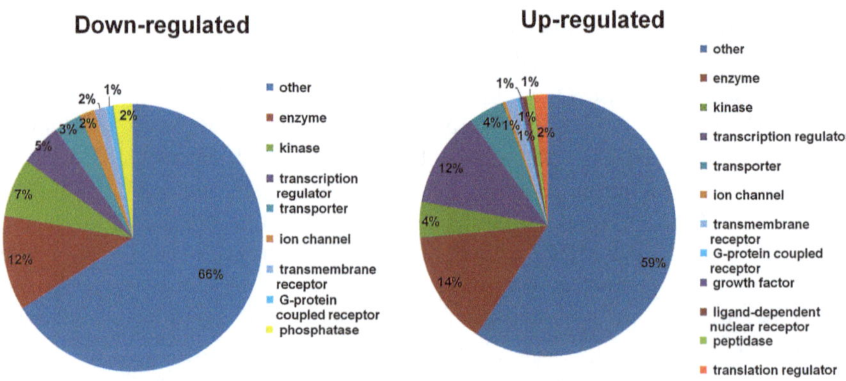

Fig. 4.11 Functional distributions of the down- (*left*) and upregulated phosphoproteins (*right*) (Reprinted with permission from Ref. [45]. Copyright 2011 American Chemical Society)

Motif-X analysis was performed to find possible kinases or kinase families contributing to the change of the protein phosphorylation level in HCC samples [46]. The prodirected motifs (PXpSP, RXXpSP, and pSP) which were mainly associated with ERK1/2 kinases were found to be overrepresented in both the upregulated and downregulated phosphorylation sites. The enrichment of substrate motifs for ERK1/2 in both the upregulated and downregulated phosphorylation sites was previously observed in the investigation of global phosphorylation changes induced by 17β-Estradiol (E2) in MCF-7 cells [47]. The extracellular signal-regulated kinases (ERKs) are best studied in the mitogen-activated protein kinase (MAPK)

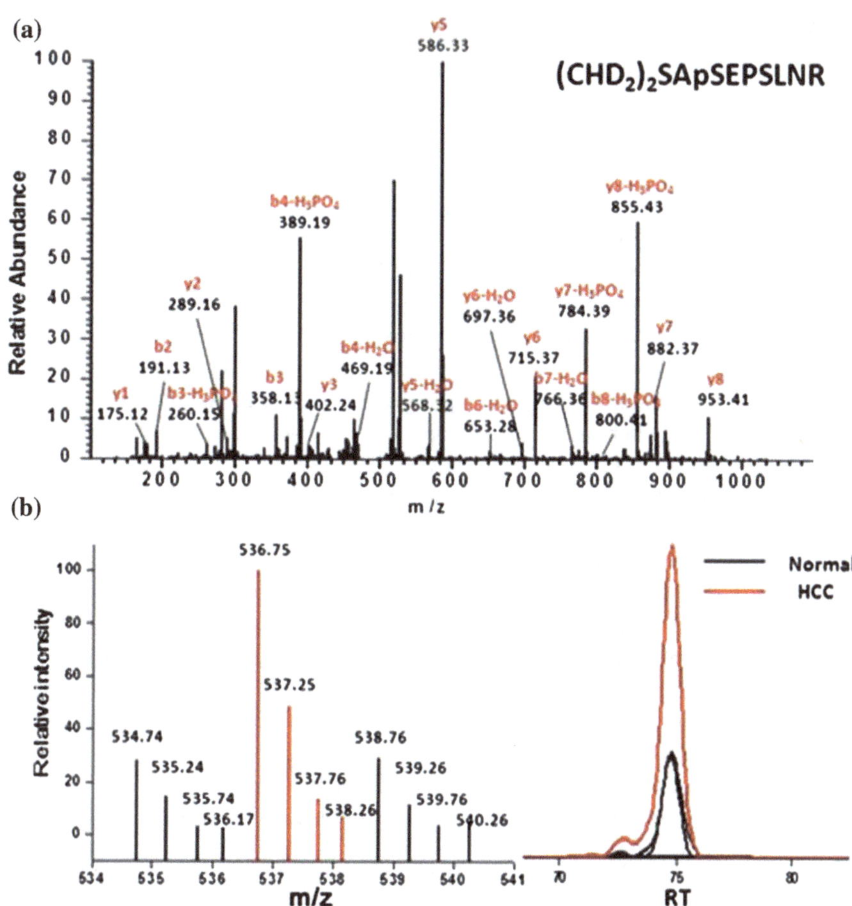

Fig. 4.12 **a** MS/MS spectrum of the phosphopeptide SApSEPSLNR from B-Raf. **b** MS spectrum and XIC peaks showed significant increase of SApSEPSLNR in HCC samples (Reprinted with permission from Ref. [45]. Copyright 2011 American Chemical Society)

pathways and active ERKs can phosphorylate many cytoplasmic and nuclear targets to regulate cell proliferation [48]. Significant overrepresentation of the ERKs has been observed in many human cancers [48, 49]. Our quantification results emphasized the roles of ERKs in the regulation of protein phosphorylation in human liver tissue during HCC progressing. For example, B-Raf is the core component in the ERK signaling cascade and also is a major contributor to numerous human cancers [50, 51]. The phosphopeptide SApSEPSLNR from B-Raf was identified with high confidence in this study (MS/MS spectra in Fig. 4.12a) and corresponding phosphorylation site S729 was upregulated by twofold. Both the intensity and XIC of these triplex isotope pairs indicated a twofold increase of this phosphopeptide as shown in Fig. 4.12b.

4.2.5 Phosphoproteome Analysis of an Early Onset Mouse Model (TgCRND8) of Alzheimer's Disease

We then applied the pseudo triplex isotope dimethyl labeling strategy to investigate the dynamic changes of phosphoproteome during the progression of the disease in an animal model of AD. Two time points (2 and 6 months) were selected for this study and two biological replicate analyses were performed at each time point. The hippocampus of two pairs of animal (Tg and NonTg) at each time point was used for comparative analysis. Briefly, for the first analysis at each time point identical amount of two nontransgenic (NonTg) mouse samples were labeled with light (L) and heavy (H) dimethyl group, respectively; while the transgenic (Tg) mouse sample was labeled with medium (M) dimethyl group. Then, the three isotopic labeling samples were mixed together and the phosphopeptides were enriched by Ti^{4+}-IMAC followed with nanoflow 2D LC-MS/MS analysis. Two replicate quantification of Tg and NonTg sample can be obtained in just one experiment (M/L and M/H) as described in previous section. In contrast, in the second biological replicate analysis, the same amount of two Tg samples were labeled with light and heavy dimethyl group, respectively; and the NonTg mouse sample was labeled with medium dimethyl group. In total, we identified 1,026 phosphopeptides representing 1,168 phosphorylation sites (83 % pS, 15 % pT, and 2 % pY) from 476 unique proteins. Of the 1,168 identified phosphorylations sites, 82 % were assigned to a protein kinases according to the PHOSIDA database, whereas 18 % had unknown phosphorylation motifs. A Bioinformatics analysis reveals that most of the phosphoproteins come from the plasma membrane, nonmembrane-bounded organelle, intracellular nonmembrane-bounded organelle, plasma membrane part, and cytoskeleton (Fig. 4.13).

595 phosphopeptides from 293 unique proteins could be confidently quantified by using MaxQuant. Of these, statistically significant changes in the ratios of 139 phosphopeptides were detected in hippocampus of the Tg mice compared to the congenic NonTg control mice upon conversion to a symptomatic state (i.e., 2 vs. 6 months of age). These data further identify the subset of phosphosignaling targets altered following symptomatic onset in Tg mice. Bioinformatics analysis of the corresponding protein functions for this phosphopeptide subset revealed that the top five functions are (i) neurological disease, (ii) cellular assembly and organization, (iii) cellular function and maintenance, (iv) cellular movement, and (v) nervous system development and function. Pathway analysis performed on the changing phosphoproteins revealed that top five pathways were enriched including: protein kinase A signalling, calcium signaling, synaptic long-term potentiation, axonal guidance signalling, and pantothenate and CoA biosynthesis. Most of these functions and pathways are related to the AD progression. For example, six phosphoproteins involved in the long-term potentiation (LTP) pathway were quantified in our experiments. The phosphorylation level of some specific phosphopeptides derived from PKA, PP1, CaMK2, PKC, and mGluR increased during the progression of the AD. However, the phosphorylation level of CaN (PPP3CA) and some sites of PKC decreased over time.

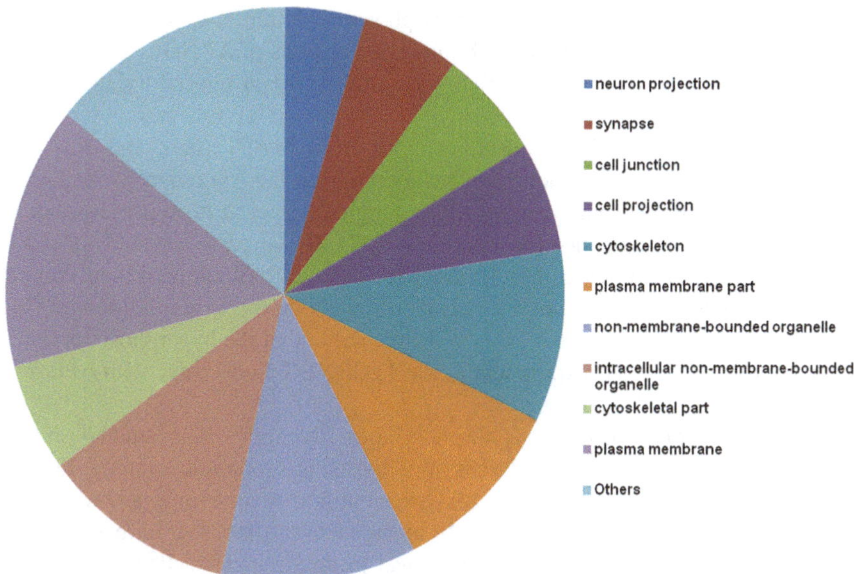

Fig. 4.13 Localization of the quantified phosphoproteins

Neuronal connections are formed by the extension of axons, which migrate to reach their synaptic targets. The axonal growth cone, located at the axon leading edge, contains receptors that sense attractive and repulsive guidance cues, which help navigate the axon to its final destination. Twenty-two phosphorylation sites, corresponding to 11 phosphoproteins within this pathway were quantified in our experiments. The phosphorylation level of ARHGEF6/7, LINGO1, NOGO, and MAG were significantly downregulated over time; but the phosphorylation level of srGAP, PKA, and some sites of CRMP2 and PKC were significantly upregulated during the progression of the AD. Interestingly, recent reports implicated the abnormal regulation of CRMP2 (Collapsin Response Mediator Protein 2) in AD pathogenesis and the development of neurofibrillary tangles whereas the increase in APP expression and proteolytic processing by PS-1 leads to CRMP2 hyperphosphorylation [52]. In our experiments, the Thr521 and Ser522 were found to be significantly upregulated over time. We also checked other isoforms of the CRMP proteins family, and it was observed that the Ser521 of CRMP1 is also significantly upregulated, but the Ser518 of CRMP1 is significantly downregulated during the disease progression.

Interestingly, the protein SHANK3, which is a large scaffold postsynaptic density protein implicated in dendritic spines and synapse formation, is significantly downregulated in phosphorylation level of Thr854 and Ser856. The level of SHANK3 was recently shown to decrease during Aβ oligomers progressive accumulation in APP transgenic mice [53]. Another recent study also reported that RNAi knockdown of SHANK3 led to impaired signaling via mGluR5, through reduction in ERK1/2 and CREB phosphorylation [54]. Furthermore, we can see that SHANK3 interacted with the gene products of ARHGEF7, DLGAP1, DLGAP2, and DLGAP3 show similar regulation trend in phosphorylation level, but GRM5 shows opposite trend [55].

We only quantified eight phosphopeptides from MAPT (tau), and only two phosphorylation sites show moderate elevation during the AD progression. This is strange as the intraneuronal neurofibrillary tangle composed of hyperphosphorylated tau is considered as one of the features of AD. However, recent study also discovered similar phenomenon that Aβ-mediated cell death of cultured hippocampal neurons reveals extensive Tau fragmentation without increased full-length Tau phosphorylation [56]. Therefore, it is understandable that the hyperphosphorylated tau is not observed in our experiment as the TgCRND8 model mice with APP over expressed are utilized.

Therefore, global proteome phosphorylation dynamics was studied during the AD progression of the TgCRND8 mice by using proteomic strategy. Finally, 1,026 phosphopeptides were successfully indentified across time, among which 139 phosphopeptides were significantly changed over time. We have systematically analyzed these datasets and some novel protein phosphorylations were discovered, such as the CRMP2 phosphorylation level was significantly elevated and SHANK3 phosphorylation level was significantly downregulated. We think these datasets could be provided as a useful public resource and give some meaningful insights for further biological studies.

4.2.6 Conclusion

Proteome quantification is extremely important to investigate the biological functions of different proteins. Stable isotope labeling strategy is the most popular way in all of the proteome quantification methods. There are still some problems in isotope labeling proteome quantification, such as the quantification sensitivity, throughput, and accuracy still need to be improved. In this chapter, we first developed an online isotope labeling system by using the RP-SCX biphasic trap column, all of the dimethyl isotope labeling procedures can be automatically performed by the auto sampler, which greatly decreases the sample loss and contaminant during the labeling process. After isotope labeling, the labeled peptides mixture is transferred to the SCX segment of the biphasic trap column and an online 2D LC-MS/MS analysis is automatically performed. As both isotope labeling and 2D LC-MS/MS analysis are online performed, the proteome quantification accuracy, sensitivity, and coverage are all improved. Then, to further improve the accuracy of proteome quantification, a pseudo triplex isotope labeling strategy was developed to realize technical replicate analyses in just one quantification experiment. After controlling the CV in technical replicate analyses, high quantification accuracy was achieved. Finally, this pseudo triplex isotope labeling strategy was successfully applied to investigate the different phosphoproteome expressions of HCC and normal liver tissues as well as the phosphoproteome changes in different time point of the progression of the AD model mice. Therefore, our new proteome quantification strategies exhibit great potential in future analyses of different types of biological samples.

4.3 Experimental Section

4.3.1 Materials

Daisogel ODS-AQ (5 μm, 12 nm pore) was purchased from DAISO Chemical CO., Ltd. (Osaka, Japan). PEEK tubing, sleeves, microtee, and microcross were obtained from Upchurch Scientific (Oak Harbor, WA, USA). Fused silica capillaries with 50- and 200-μm i.d. were purchased from Yongnian Optical Fiber Factory (Hebei, China), and with 75-μm i.d. from Polymicro Technologies (Phoenix, AZ, USA). All the water used in experiments was purified using a Mill-Q system from Millipore Company (Bedford, MA, USA). Ethylene glycol methacrylate phosphate (EGMP), γ-methacryloxypropyltrimethoxysilane (γ-MAPS), methylene bis-acrylamide, formaldehyde, sodium cyanoborohydride, dithiothreitol (DTT), iodoacetamide, and trypsin were obtained from Sigma (St. Louis, MO, USA). Azobisisobutyronitrile (AIBN) was obtained from Shanghai Fourth Reagent Plant (Shanghai, China). Formic acid (FA) was obtained from Fluka (Buches, Germany). Acetonitrile (ACN, HPLC grade) was from Merck (Darmstadt, Germany).

4.3.2 Sample Preparation

The HCC and normal human liver tissues were obtained from the Second Affiliated Hospital of Dalian Medical University (Dalian, China). The utilization of human tissues complied with guideline of Ethics Committee of the Hospital. The normal liver tissues have been verified by histopathological examination, which excluded the presence of invading or microscopic metastatic cancer cells. The protein extraction procedures were the same as our previous works [57, 58]. Briefly, human liver tissue was placed in an ice-cold homogenization buffer consisting of 8 M urea, 4 % CHAPS w/v, 65 mM DTT, a mixture of protease inhibitor (Complete Mini protease inhibitor cocktail tablets, one tablet for 10 mL homogenization buffer), and 100 mM NH_4HCO_3 at pH 7.8. After homogenized by a Potter-Elvehjem homogenizer with a Teflon piston, sonicated for 100 W × 30 s, and centrifuged at 25,000 g for 1 h, the protein concentration was determined by Bradford assay. In our study, the same amount of protein extracts from five HCC or normal liver tissues were mixed together, respectively. Then, the proteins in the

HCC or normal samples were precipitated by chloroform/methanol precipitation as described by Wessel et al. [59]. After washing with methanol, the pellets were resuspended in 1 mL denaturing buffer containing 50 mM Tris/HCl (pH 8.1) and 8 M urea and the protein concentration was determined again by Bradford assay. The protein samples were reduced by DTT at 37 °C for 2 h and alkylated by iodoacetamide in dark at room temperature for 40 min. Then the solutions were diluted to 1 M urea with 50 mM Tris/HCl (pH 8.1). Finally, trypsin was added with weight ratio of trypsin to protein at 1/25 and incubated at 37 °C overnight. Then, the tryptic digests were purified with a homemade C18 solid phase cartridge and exchanged into buffer A (0.1 % formic acid water solution). Finally, the samples were stored at −30 °C before usage.

TgCRND8 mice with a double mutant form of APP695 (KM670/671NL + V717F) and wild-type litter-mates controls were used. Eight female mice (4 TgCRND8 and wild-type litter-mates controls were used. Eight female mice (4 TgCRND8 and four WT) were sacrificed at 2, 4, and 6 months of age. Two animals of each strain per time point were used for proteome quantification while the remaining animals were used for western blot validation. The hippocampus were dissected and quick-frozen in liquid nitrogen before storage at −80 °C. All animal manipulations were performed according to the ethical guidelines for experimentation established by the Canadian Council for Animal Care. The protein extraction procedures were similar to liver tissues as described above. For dimethyl isotope labeling for the phosphoproteome quantification AD mice, the protein digest was loaded onto a prewashed 200 mg tC18 cartridge (Waters) and equilibrated with 1 mL 50 mM sodium phosphate (pH 7.5). Then, 5 mL isotopic labeling reagents containing 0.2 % CH_2O and 30 mM $NaBH_3CN$ (light form labeling); or 0.2 % CD_2O and 30 mM $NaBH_3CN$ (intermediate form labeling); or 0.2 % $^{13}CD_2O$ and 30 mM $NaBD_3CN$ (heavy form labeling) was added and passed through the tC18 cartridge. 21 After equilibrating with 1 mL 0.1 % formic acid aqueous solution, the isotopic labeled peptides were eluted from the tC18 cartridge using1 mL 80 % Acetonitrile (ACN), 0.1 % formic acid aqueous solution. The eluted solution was lyophilized by Speed Vacuum (Thermo) and stored at −80 °C before usage.

For phosphopeptides enrichment, the procedures for preparation of the Ti^{4+}-IMAC monodisperse microspheres were described previously [60]. The Ti^{4+}-IMAC microspheres were well dispersed into 80 % ACN, 6 % TFA aqueous solution at a concentration of 10 mg/mL. For comparative analyses, the same amount of isotopic dimethyl labeled samples were dissolved into 0.1 % FA aqueous solution and mixed. Then, the labeled peptide mixture was incubated with 10 times Ti^{4+}-IMAC microspheres (w/w). Following 30 min incubation at 4 °C, the Ti^{4+}-IMAC microspheres with the adsorbed phosphopeptides were collected by centrifugation at 22,000 g for 15 min. The microspheres were then washed with 500 μL 50 % ACN, 6 % TFA and 200 mM NaCl aqueous solution, and 30 % ACN and 0.1 % TFA aqueous solution to remove unspecific adsorption. Finally, 500 μL $NH_3 \cdot H_2O$ (12.5 %) was used to elute the enriched phosphopeptides from the microspheres by stirring for 15 min at 4 °C and sonicating for 15 min with ice. After centrifugation at 22,000 g for 15 min, the phosphopeptides containing supernatant was collected and lyophilized by Speed Vacuum.

4.3.3 Column Preparation

To prepare the RP-SCX biphasic column, the 200-μm i.d. capillary column was pretreated with γ-MAPS as described in previous sections. 25-cm-long pretreated capillary column was filled with 10-cm-long SCX monolithic polymerization mixture containing 80 μL ethylene glycol methacrylate phosphate, 60 mg methylene bis-acrylamide, 270 μL dimethylsulfoxide, 200 μL dodecanol, 50 μL N, N'-dimethylformamide, and 2 mg AIBN. Then, the capillary column was sealed at both ends with rubber stoppers, submerged into a water bath, and allowed to react for 12 h at 60 °C. The resultant monolithic capillary column was washed with methanol using a HPLC pump to remove unreacted monomers and porogens and dried with nitrogen gas. Finally, the SCX monolithic segment was cut at 7 cm length, followed with packing C18 material into the empty capillary to reach a length of about 7 cm adjacent the SCX monolith by a homemade pneumatic pressure cell at constant nitrogen gas pressure of about 580 psi with a slurry packing method.

To prepare the separation column, one end of a 75 μm i.d. capillary was first manually pulled to a fine point of ~3 μm with a flame torch, and then the C18 particles were packed until the packing section reached the length of 15 cm by the same method as described above.

4.3.4 RP Gradient Nanoflow LC Separation

The HPLC system (Thermo, San Jose, CA) consisted of a degasser and a quaternary Surveyor MS pump. 0.1 % FA aqueous solution (buffer A) and ACN with 0.1 % formic acid (buffer B) were used for gradient separation. In all gradient separation, the flow rate after splitting was adjusted to ~300 nL/min (dashed line mode in Fig. 4.1). Two types of RP separation gradient were used in our study. The first one was 95-min gradient, which was developed from 0 to 10 % buffer B for 5 min, from 10 to 35 % for 90 min, and from 35 to 80 % for 5 min. After flushing 80 % buffer B for 10 min, the separation system was equilibrated by buffer A for 15 min. The second one was 155-min gradient, which was the same as the 95-min gradient except for from 10 to 35 % buffer B for 150 min.

4.3.5 Mass Spectrometry Analysis

The MS analysis was performed on LTQ-Orbitrap mass spectrometer (Thermo, San Jose, CA) at a resolution of 60,000. The temperature of the ion transfer capillary was set at 200 °C. The spray voltage was set at 1.8 kV and the normalized collision energy was set at 35.0 %. One microscan was set for each MS and MS/MS scan. All MS and MS/MS spectra were acquired in the data-dependent mode. Survey full scan MS was acquired from m/z 400 to 2,000, and seven most intense ions were selected for MS/MS scan by collision-induced dissociation. The target ion setting was 5e5 for the Orbitrap, with a maximum fill-time of 500 ms. MS/MS scans were acquired in the LTQ with a target ion setting of 3e4 and a maximum fill-time of 100 ms. The dynamic exclusion function was set as follows: repeat count 2, repeat duration 30 s, and exclusion duration 90 s.

4.3.6 Sequential Isotope Labeling and Online Multidimensional Separation

The configuration for automated sample injection using biphasic trap column was consisted with an autosampler, a 6-port/two-position switching valve, a microtee, and a microcross as shown in Fig. 4.1. During sample or labeling reagent injection, the switching valve was switched to close the splitting flow and the flow through from the trap column was switched to waste (yellow line mode in Fig. 4.1). At first, 20 μL tryptic digest of the first sample was injected at a flow rate of 6 μL/min for 15 min. Second, 50 μL light isotope labeling reagent containing 0.04 % formaldehyde, 6 mM cyanoborohydride, and 50 mM sodium phosphate (pH 7.5) was injected at a flow rate of 6 μL/min for 20 min. Third, 20 μL tryptic digest of the second sample was injected at a flow rate of 6 μL/min for 15 min. Fourth, 50 μL heavy isotope labeling reagent containing 0.04 % deuterated formaldehyde, 6 mM cyanoborohydride, and 50 mM sodium phosphate (pH 7.5) was injected at a flow rate of 6 μL/min for 20 min. In all the injection procedures, 0.1 % FA aqueous solvent (buffer A) was used as injection solvent, and 8 min additional equilibrium by buffer A was already included in each procedure. Then, the switching valve was switched to open the splitting flow (green line mode in Fig. 4.1).

For 1D separation, the peptides enriched onto the RP segment of the biphasic trap column was transferred to SCX segment with elution by 80 % buffer B for 10 min, followed to separation column with elution by 100 % buffer C (1,000 mM NH_4Ac, pH 2.8) for 20 min. After re-equilibrated for 15 min with buffer A, a 95-min RP gradient μLC-MS/MS analysis on separation column was conducted.

For 29-h online multidimensional separation, a 95-min RP gradient μLC-MS/MS was applied at first (0 mM fraction). Therefore, all of the peptides enriched onto the RP segment of the biphasic trap column were transferred to the SCX monolith segment, and some unretained peptides were separated by the 15-cm-long separation column and detected by MS. Then, a series of stepwise elution (generated by buffer A and buffer C) with salt concentrations of 50, 100, 150, 200, 250, 300, 350, 400, 500, and 1,000 mM NH_4Ac was used to gradually elute peptides from SCX segment onto the C18 separation column. Each salt step lasts 10 min except last one for 20 min. After whole system was re-equilibrated for 15 min with buffer A, the 95-min binary RP gradient nanoflow LC-MS/MS analysis as described above was applied to separate peptides prior to MS detection in each cycle.

To increase the proteome coverage, a 63-h online multidimensional separation was also applied in our study. The stepwise salt elution was developed as

50, 75, 100, 125, 150, 175, 200, 225, 250, 275, 300, 325, 350, 375, 400, 500, and 1,000 mM. And the 155-min separation gradient μLC-MS/MS analysis as described above was applied in each cycle. The other procedures were the same as the 29-h online multidimensional separation.

4.3.7 Protein Identification and Quantification

All the MS/MS spectra in one acquired raw file were converted to single *.mgf file using DTASupercharge (v2.0a7) [61]. Then the *.mgf file was searched against the International Protein Index (IPI) human database (v3.52, 73,928 entries) using Mascot Version 2.1 (Matrix Science). To evaluate the false discovery rate (FDR), reversed sequences were appended to the database. Cysteine residues were searched as static modification of +57.0,215 Da, methionine residues as variable modification of +15.9,949 Da. And both light and heavy dimethylation of peptide amino termini and lysine residues were set as variable modification of +28.0313 and +32.0564 Da, respectively. Peptides were searched using fully tryptic cleavage constraints and up to two missed cleavages sites were allowed for tryptic digestion. The mass tolerances were 10 ppm for parent masses and 0.8 Da for fragment masses. Peptides with Mascot score ≥25 (rank 1, $P \leq 0.05$) were used for protein quantification.

Protein quantification was performed using a dimethyl-adapted version of MSQuant (v2.0a81). Peptide ratios were obtained by calculating the extracted ion chromatograms (XIC) of the light and heavy forms of the peptide using the monoisotopic peaks only and protein ratios were calculated from the average of the all quantified peptides [61–63]. Then, all the MSQuant outputs of the same online multidimensional separation were imported in StatQuant (v1.2.2) and the quantified proteins were normalized against the log2 of the median ratio of all peptides quantified [62, 64]. In pseudo triplex isotope labeling or conventional replicate proteome quantification, CV < 50 % or 40 % were applied to control the peptides with two times change of relative ratio less than 1 % for standard protein sample quantification.

For phosphoproteome analysis, the *.raw files obtained from the LTQ-OrbiTrap were directly processed by Maxquant (http://maxquant.org/, version 1.1.1.36) to identify and quantify the phosphopeptides [17, 65, 66]. The database searching and data filtering criterion was similar as described above.

References

1. Aebersold R, Mann M (2003) Mass spectrometry-based proteomics. Nature 422:198–207
2. Mann M (2006) Functional and quantitative proteomics using SILAC. Nat Rev Mol Cell Biol 7:952–958
3. Gruhler A, Olsen JV, Mohammed S, Mortensen P, Faergeman NJ, Mann M, Jensen ON (2005) Quantitative phosphoproteomics applied to the yeast pheromone signaling pathway. Mol Cell Proteomics 4:310–327

4. Forner F, Foster LJ, Campanaro S, Valle G, Mann M (2006) Quantitative proteomic comparison of rat mitochondria from muscle, heart, and liver. Mol Cell Proteomics 5:608–619
5. Wright ME, Han DK, Aebersold R (2005) Mass spectrometry-based expression profiling of clinical prostate cancer. Mol Cell Proteomics 4:545–554
6. Kruger M, Moser M, Ussar S, Thievessen I, Luber CA, Forner F, Schmidt S, Zanivan S, Fassler R, Mann M (2008) SILAC mouse for quantitative proteomics uncovers kindlin-3 as an essential factor for red blood cell function. Cell 134:353–364
7. Ong S-E, Mann M (2005) Mass spectrometry-based proteomics turns quantitative. Nat Chem Biol 1:252–262
8. Khoury GA, Baliban RC, Floudas CA (2011) Proteome-wide post-translational modification statistics: frequency analysis and curation of the swiss-prot database. Sci Rep 1
9. Mann M, Jensen ON (2003) Proteomic analysis of post-translational modifications. Nat Biotechnol 21:255–261
10. Witze ES, Old WM, Resing KA, Ahn NG (2007) Mapping protein post-translational modifications with mass spectrometry. Nat Methods 4:798–806
11. Wang W, Zhou H, Lin H, Roy S, Shaler TA, Hill LR, Norton S, Kumar P, Anderle M, Becker CH (2003) Quantification of proteins and metabolites by mass spectrometry without isotopic labeling or spiked standards. Anal Chem 75:4818–4826
12. Radulovic D, Jelveh S, Ryu S, Hamilton TG, Foss E, Mao Y, Emili A (2004) Informatics platform for global proteomic profiling and biomarker discovery using liquid chromatography-tandem mass spectrometry. Mol Cell Proteomics 3:984–997
13. Gygi SP, Rist B, Gerber SA, Turecek F, Gelb MH, Aebersold R (1999) Quantitative analysis of complex protein mixtures using isotope-coded affinity tags. Nat Biotechnol 17:994–999
14. Han DK, Eng J, Zhou H, Aebersold R (2001) Quantitative profiling of differentiation-induced microsomal proteins using isotope-coded affinity tags and mass spectrometry. Nat Biotechnol 19:946–951
15. Kettenbach AN, Rush J, Gerber SA (2011) Absolute quantification of protein and post-translational modification abundance with stable isotope-labeled synthetic peptides. Nat Protocols 6:175–186
16. Ong S-E, Mann M (2007) A practical recipe for stable isotope labeling by amino acids in cell culture (SILAC). Nat Protocols 1:2650–2660
17. Cox J, Matic I, Hilger M, Nagaraj N, Selbach M, Olsen JV, Mann M (2009) A practical guide to the MaxQuant computational platform for SILAC-based quantitative proteomics. Nat Protocols 4:698–705
18. Cox J, Mann M (2008) MaxQuant enables high peptide identification rates, individualized p.p.b.-range mass accuracies and proteome-wide protein quantification. Nat Biotechnol 26:1367–1372
19. Boersema PJ, Raijmakers R, Lemeer S, Mohammed S, Heck AJR (2009) Multiplex peptide stable isotope dimethyl labeling for quantitative proteomics. Nat Protocols 4:484–494
20. Hsu J-L, Huang S-Y, Chow N-H, Chen S-H (2003) Stable-isotope dimethyl labeling for quantitative proteomics. Anal Chem 75:6843–6852
21. Huang S-Y, Tsai M-L, Wu C-J, Hsu J-L, Ho S-H, Chen S-H (2006) Quantitation of protein phosphorylation in pregnant rat uteri using stable isotope dimethyl labeling coupled with IMAC. Proteomics 6:1722–1734
22. Boersema PJ, Aye TT, van Veen TAB, Heck AJR, Mohammed S (2008) Triplex protein quantification based on stable isotope labeling by peptide dimethylation applied to cell and tissue lysates. Proteomics 8:4624–4632
23. Raijmakers R, Berkers CR, de Jong A, Ovaa H, Heck AJR, Mohammed S (2008) Automated online sequential isotope labeling for protein quantitation applied to proteasome tissue-specific diversity. Mol Cell Proteomics 7:1755–1762
24. Seow TK, Liang RC, Leow CK, Chung MC (2001) Hepatocellular carcinoma: from bedside to proteomics. Proteomics 1:1249–1263
25. Zheng J, Gao X, Beretta L, He F (2006) The Human Liver Proteome Project (HLPP) workshop during the 4th HUPO World Congress. Proteomics 6:1716–1718
26. Chaerkady R, Harsha HC, Nalli A et al (2008) A quantitative proteomic approach for identification of potential biomarkers in hepatocellular carcinoma. J Proteome Res 7:4289–4298

27. Chen N, Sun W, Deng X et al (2008) Quantitative proteome analysis of HCC cell lines with different metastatic potentials by SILAC. Proteomics 8:5108–5118
28. Selkoe DJ (2001) Alzheimer's disease: genes, proteins, and therapy. Physiol Rev 81:741–766
29. Blennow K, de Leon MJ, Zetterberg H (2006) Alzheimer's disease. Lancet 368:387–403
30. Rhein V, Song X, Wiesner A et al (2009) Amyloid-beta and tau synergistically impair the oxidative phosphorylation system in triple transgenic Alzheimer's disease mice. Proc Natl Acad Sci U S A 106:20057–20062
31. Dierssen M, Fillat C, Crnic L, Arbones M, Florez J, Estivill X (2001) Murine models for Down syndrome. Physiol Behav 73:859–871
32. David DC, Ittner LM, Gehrig P, Nergenau D, Shepherd C, Halliday G, Gotz J (2006) Beta-amyloid treatment of two complementary P301L tau-expressing Alzheimer's disease models reveals similar deregulated cellular processes. Proteomics 6:6566–6577
33. Gillardon F, Rist W, Kussmaul L, Vogel J, Berg M, Danzer K, Kraut N, Hengerer B (2007) Proteomic and functional alterations in brain mitochondria from Tg2576 mice occur before amyloid plaque deposition. Proteomics 7:605–616
34. Chishti MA, Yang DS, Janus C et al (2001) Early-onset amyloid deposition and cognitive deficits in transgenic mice expressing a double mutant form of amyloid precursor protein 695. J Biol Chem 276:21562–21570
35. Phinney AL, Drisaldi B, Schmidt SD et al (2003) In vivo reduction of amyloid-beta by a mutant copper transporter. Proc Natl Acad Sci U S A 100:14193–14198
36. Ryan SD, Whitehead SN, Swayne LA et al (2009) Amyloid-beta42 signals tau hyperphosphorylation and compromises neuronal viability by disrupting alkylacylglycerophosphocholine metabolism. Proc Natl Acad Sci U S A 106:20936–20941
37. Hawkes CA, McLaurin J (2009) Selective targeting of perivascular macrophages for clearance of beta-amyloid in cerebral amyloid angiopathy. Proc Natl Acad Sci U S A 106:1261–1266
38. Wang F, Chen R, Zhu J, Sun D, Song C, Wu Y, Ye M, Wang L, Zou H (2010) A fully automated system with online sample loading, isotope dimethyl labeling and multidimensional separation for high-throughput quantitative proteome analysis. Anal Chem 82:3007–3015
39. Neo SY, Leow CK, Vega VB, Long PM, Islam AFM, Lai PBS, Liu ET, Ren EC (2004) Identification of discriminators of hepatoma by gene expression profiling using a minimal dataset approach. Hepatology 39:944–953
40. Liang CRMY, Leow CK, Neo JCH, Tan GS, Lo SL, Lim JWE, Seow TK, Lai PBS, Chung MCM (2005) Proteome analysis of human hepatocellular carcinoma tissues by two-dimensional difference gel electrophoresis and mass spectrometry. Proteomics 5:2258–2271
41. Liang RCMY, Neo JCH, Lo SL, Tan GS, Seow TK, Chung MCM (2002) Proteome database of hepatocellular carcinoma. J Chromatogr B 771:303–328
42. Lemeer S, Jopling C, Gouw J, Mohammed S, Heck AJ, Slijper M, den Hertog J (2008) Comparative phosphoproteomics of zebrafish Fyn/Yes morpholino knockdown embryos. Mol Cell Proteomics 7:2176–2187
43. Kuramitsu Y, Harada T, Takashima M et al (2006) Increased expression and phosphorylation of liver glutamine synthetase in well-differentiated hepatocellular carcinoma tissues from patients infected with hepatitis C virus. Electrophoresis 27:1651–1658
44. Lee HJ, Na K, Kwon MS, Kim H, Kim KS, Paik YK (2009) Quantitative analysis of phosphopeptides in search of the disease biomarker from the hepatocellular carcinoma specimen. Proteomics 9:3395–3408
45. Song C, Wang F, Ye M, Cheng K, Chen R, Zhu J, Tan Y, Wang H, Figeys D, Zou H (2011) Improvement of the quantification accuracy and throughput for phosphoproteome analysis by a pseudo triplex stable isotope dimethyl labeling approach. Anal Chem 83:7755–7762
46. Schwartz D, Gygi SP (2005) An iterative statistical approach to the identification of protein phosphorylation motifs from large-scale data sets. Nat Biotechnol 23:1391–1398
47. Wu CJ, Chen YW, Tai JH, Chen SH (2011) Quantitative phosphoproteomics studies using stable isotope dimethyl labeling coupled with IMAC-HILIC-nanoLC-MS/MS for estrogen-induced transcriptional regulation. J Proteome Res 10:1088–1097

48. Dhillon AS, Hagan S, Rath O, Kolch W (2007) MAP kinase signalling pathways in cancer. Oncogene 26:3279–3290
49. Reddy KB, Nabha SM, Atanaskova N (2003) Role of MAP kinase in tumor progression and invasion. Cancer Metastasis Rev 22:395–403
50. Dhomen N, Marais R (2007) New insight into BRAF mutations in cancer. Curr Opin Genet Dev 17:31–39
51. Ritt DA, Monson DM, Specht SI, Morrison DK (2010) Impact of feedback phosphorylation and Raf heterodimerization on normal and mutant B-Raf signaling. Mol Cell Biol 30:806–819
52. Soutar MP, Thornhill P, Cole AR, Sutherland C (2009) Increased CRMP2 phosphorylation is observed in Alzheimer's disease; does this tell us anything about disease development? Curr Alzheimer Res 6:269–278
53. Pham E, Crews L, Ubhi K et al (2010) Progressive accumulation of amyloid-beta oligomers in Alzheimer's disease and in amyloid precursor protein transgenic mice is accompanied by selective alterations in synaptic scaffold proteins. FEBS J 277:3051–3067
54. Verpelli C, Dvoretskova E, Vicidomini C et al (2011) Importance of shank3 in regulating metabotropic glutamate receptor 5 (mGluR5) expression and signaling at synapses. J Biol Chem 286(40):34839–34850
55. Grabrucker AM, Schmeisser MJ, Udvardi PT et al (2011) Amyloid beta protein-induced zinc sequestration leads to synaptic loss via dysregulation of the ProSAP2/Shank3 scaffold. Mol Neurodegener 6:65
56. Reifert J, Hartung-Cranston D, Feinstein SC (2011) Amyloid beta-mediated cell death of cultured hippocampal neurons reveals extensive Tau fragmentation without increased full-length tau phosphorylation. J Biol Chem 286:20797–20811
57. Han GH, Ye ML, Zhou HJ, Jiang XN, Feng S, Jiang XG, Tian RJ, Wan DF, Zou HF, Gu JR (2008) Large-scale phosphoproteome analysis of human liver tissue by enrichment and fractionation of phosphopeptides with strong anion exchange chromatography. Proteomics 8:1346–1361
58. Chen R, Jiang XN, Sun DG, Han GH, Wang FJ, Ye ML, Wang LM, Zou HF (2009) Glycoproteomics analysis of human liver tissue by combination of multiple enzyme digestion and hydrazide chemistry. J Proteome Res 8:651–661
59. Wessel D, Flügge UI (1984) A method for the quantitative recovery of protein in dilute solution in the presence of detergents and lipids. Anal Biochem 138:141–143
60. Yu Z, Han G, Sun S, Jiang X, Chen R, Wang F, Ye M, Zou H (2009) Preparation of monodisperse immobilized Ti4+ affinity chromatography microspheres for specific enrichment of phosphopeptides. Anal Chim Acta 636:34–41
61. Mortensen P, Gouw JW, Olsen JV et al (2009) MSQuant, an open source platform for mass spectrometry-based quantitative proteomics. J Proteome Res 9(1):393–403
62. Lemeer S, Jopling C, Gouw J, Mohammed S, Heck AJR, Slijper M, den Hertog J (2008) Comparative phosphoproteomics of zebrafish Fyn/Yes morpholino knockdown embryos. Mol Cell Proteomics 7:2176–2187
63. Chen N, Sun W, Deng XY et al (2008) Quantitative proteome analysis of HCC cell lines with different metastatic potentials by SILAC. Proteomics 8:5108–5118
64. van Breukelen B, van den Toorn HWP, Drugan MM, Heck AJR (2009) StatQuant: a post-quantification analysis toolbox for improving quantitative mass spectrometry. Bioinformatics 25:1472–1473
65. Cox J, Neuhauser N, Michalski A, Scheltema RA, Olsen JV, Mann M (2011) Andromeda: a peptide search engine integrated into the MaxQuant environment. J Proteome Res 10(4):1794–1805
66. Cox J, Mann M (2008) MaxQuant enables high peptide identification rates, individualized p.p.b.-range mass accuracies and proteome-wide protein quantification. Nat Biotechnol 26:1367–1372

Zeitfracht Medien GmbH
Ferdinand-Jühlke-Straße 7
99095 Erfurt, Deutschland
produktsicherheit@kolibri360.de